P9-DTY-560

Maintainability Engineering

Maintainability Engineering

David J Smith
BSc(Eng) CEng MIEE MITO

and

Alex H Babb
BSc(Eng) CEng MIEE ACGI

A Halsted Press Book

JOHN WILEY AND SONS, NEW YORK

First published in Great Britain by Sir Isaac Pitman & Sons Ltd 1973

Published in the U.S.A. by Halsted Press
a division of
John Wiley & Sons Inc.,
New York

Library of Congress Cataloging in Publication Data

Smith, David John.
Maintainability engineering.

"A Halsted Press book."
Bibliography: p.
1. Maintainability (Engineering) I. Babb, Alex
H., joint author. II. Title.

TS174.S64 620'.0046 73-5368

ISBN 0-470-80199-9

Library of Congress Catalog Card No: 73-5368

Made in Great Britain

Preface

It is often said that the achievement of maintainability as well as reliability objectives is largely a matter of good engineering practice. It is best, however, when setting out to achieve specific engineering goals to formulate and adhere to a planned sequence of activities together with periodic reviews of progress. Such a programme is essential where a high order of system effectiveness is required and can then be geared to the required objectives in the light of the particular product, its operating conditions, maintenance support limitations, outage costs and many other relevant factors. The programme will include activities, often described as good engineering, which are essential to any development project. In addition to these, however, there will be activities such as specific tests and reviews dictated by the existence of specific maintainability and reliability objectives. The term *system effectiveness* has been defined in many ways, but in this book the authors have used it as a collective description of maintainability, reliability, availability, human factors and safety.

The purpose of the book is to help to identify these essential activities; to provide a checklist of design and maintenance support features which influence repair times; to discuss and describe methods of evaluating the maintainability of proposed designs; and to highlight the pitfalls inherent in a contractual agreement concerning maintainability and reliability objectives.

The book refers largely to electronic equipment with a small mechanical content but the principles discussed nevertheless have a wider application.

Occasional reference to mathematics is made where relevant, but it should be emphasized that maintainability objectives can be achieved with little mathematical accompaniment.

The authors are indebted to the International Telephone and Telegraph Corporation for permission to make use of their failure report form, to the British Civil Aviation Authority for permission to use the National Air Traffic Control service report forms, and to the United States Naval Air Systems Command for permission to quote extensively from US Military Handbooks 470, 471 and 472. The authors are also indebted to Standard Telephones and Cables Ltd for permission to make use of illustrations and a description of instrument landing system equipment for the prediction exercise in Chapter 6. We are also grateful to colleagues, too numerous to mention, whose contributions to maintainability studies have influenced this book.

D.J.S.
A.H.B.

Contents

Introduction

With the steady increase in complexity of equipment, in the stringency of operating conditions, and in the positive identification of system effectiveness requirements, it is becoming harder to satisfy the requirements, and more and more emphasis is placed on preventive maintenance together with the speedy repair of replicated units as a means of achievement. Test equipment is also becoming more and more complex, and since labour and transport costs have risen dramatically in post-war years, both preventive and corrective maintenance costs represent a major part of system operating costs.

The larger and more complex a system, the greater is the capital investment it will represent and the greater its likely revenue-earning capacity. Each minute out of service is therefore going to result in considerable financial loss to the system user. The outage cost of a high-capacity telephone or data transmission link is measured in thousands of pounds per hour.

For some years in the United States and more recently in the United Kingdom it has been the practice of purchasers of complex systems to demand contracts involving stated reliability objectives either to be proved as a condition of acceptance or type approval and in some cases with financial penalties to the supplier for failures in service. In the USA similar contractual statements of repair time objectives are often included in such contracts with suppliers called upon to demonstrate the ability of their equipment to meet these objectives. This state of affairs requires a good customer/supplier liaison and understanding due to the way in which maintainability, as well as reliability, is actually influenced by both parties.

The factors which determine the maintainability of a system fall under three main headings:

Design, e.g. alarms, test points, access, etc.
Human factors, e.g. skills, training, etc.
Maintenance environment, e.g. tools, logistics, etc.

These factors will be examined in more detail in later chapters. The effect of one of these factors can seldom be determined without considering the total maintenance picture since the repair times are usually dependent upon interacting combinations of factors.

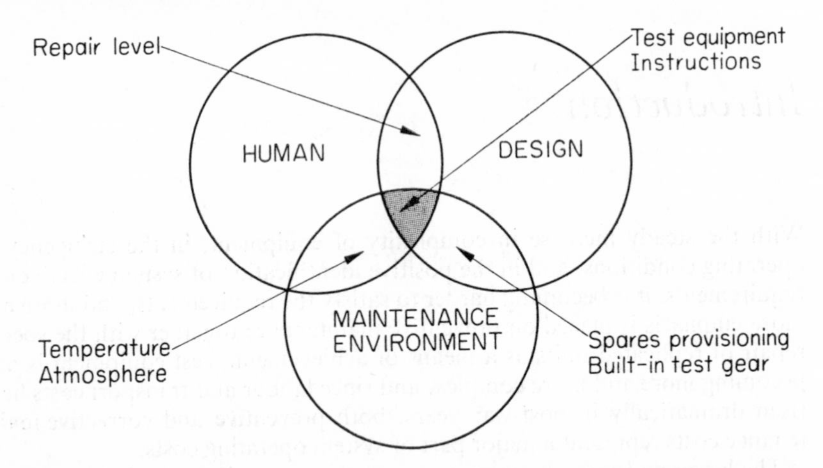

Maintainability engineering is therefore an attempt to achieve some repair time objectives by specifying a combination of design and human factors together with the maintenance philosophy consistent with the other engineering and cost constraints which exist. This involves electrical and mechanical engineering, ergonomics and human and industrial relations. It is likely therefore to appeal to the "multi-discipline" engineer who is used to taking the broad view of problems and who can think in terms of a total system and the need to optimize many interacting variables.

I

Maintainability—A System Parameter

A system is designed to achieve a given performance and its quality is the degree to which it meets this performance specification. Performance is normally specified in terms of the acceptable limits of such parameters as the maximum permissible noise or the minimum acceptable output power of a speech transmission channel, the maximum number of lost calls or the maximum switching time in a telephone exchange, the stability of a power supply or the frequency limits of an oscillator, and so on. It follows that system failure is defined as a departure from these specified limits. Failure may be defined at many levels whereas only a system failure gives rise to complete loss of system use. A unit or sub-system failure may or may not give rise to a system failure depending upon the presence, or otherwise, of redundancy. Redundancy is a design configuration, as in the duplication or triplication of units within a control system, whereby the failure of some part of the system does not result in a system failure. Reliability is frequently enhanced by the use of redundancy, but the total number of unit failures requiring repair, hence the amount of maintenance, is usually increased due to the additional equipment.

The time (hence cost) of carrying out a repair, or preventive maintenance action, at system and unit levels is therefore of interest to the customer. Given that a product initially meets the specification (i.e. it is of the required quality) then the magnitude of the following are of interest to the customer:

Cost of routine maintenance (preventive maintenance)
Cost of repair (corrective maintenance)
Cost of system outage or degraded service (loss of revenue)

The above factors are determined by the frequency of failure, the time to repair, the cost of manpower and maintenance equipment, the diversity, quantity and cost of spares carried, transportation of manpower and spares and the degree of skill required of the maintenance personnel. The first two of these (frequency of failure and time to repair) are parameters of reliability and maintainability respectively and are defined as follows.

Maintainability is the probability that a device that has failed will be restored to operational effectiveness within a given period of time when the maintenance action is performed in accordance with prescribed procedures. This is usually stated in terms of its parameter *mean time to repair*, MTTR, sometimes expressed as the *repair rate*, μ, equal to 1/MTTR. The word "repair" implies that we are concerned with the time to perform corrective maintenance actions only, whereas the time taken to carry out preventive maintenance is also of interest. The factors giving rise to an improvement in corrective repair time will, needless to say, reduce preventive maintenance time in the majority of cases.

Reliability, on the other hand, is the probability that a system will operate for some determined period of time, under the working conditions for which it was designed. This is expressed in terms of the *failure rate*, $\lambda(t)$, or in terms of the *mean time between failures*, MTBF, both of which are parameters of reliability. It can be shown that if failures occur at random and the failure rate, $\lambda(t)$, is therefore a constant, λ, the reliability, $R(t)$ is equal to $e^{-\lambda t}$. In this case t is the period of time referred to in the definition of reliability. It can also be shown that the MTBF, θ, can be obtained from the expression

$$\theta = \int_0^\infty R(t)\, dt$$

The choice of parameter (reliability, MTBF, or failure rate) depends on a number of factors, but it is not the object of this book to provide a study of reliability.*

Maintainability must not be confused with maintenance, since the former is a design parameter concerned with the minimization of repair time, whereas maintenance is the action of carrying out a repair or servicing of equipment. The way in which maintenance is carried out may influence the maintainability of a system, since the skill of maintenance personnel, the effectiveness of maintenance instructions, the quantity and type of built-in test equipment, the location of test points and many other factors influence the repair time. Furthermore, both maintainability and maintenance play a large part in the reliability of a system, since preventive maintenance can reduce the incidence of failures, and faster repair of redundant units reduces the incidence of system failures as will be discussed in Chapter 4.

Both maintainability and reliability which, as has been pointed out, affect each other, influence the costs of routine maintenance, repair and system outage. The parameter, *availability*, which combines maintainability and reliability is therefore of interest. One definition of availability, known as the steady-state definition, is that it is the proportion of time during which a system is available for use (i.e. not in a failed state). This can be shown to be $\theta/(\theta + \text{MTTR})$.

* See Smith, D. J., *Reliability Engineering* (Pitman, 1972).

Reliability is achieved by means of a combination of the following: redundancy, component derating, use of reliable components, limiting complexity, speedy repair of redundant units, removing early failures by means of burn-in, worst-case design (to reduce sensitivity to variations of parameters or operating conditions), process control and quality assurance. The following activities are necessary for a reliability programme: design review, parts selection and control, failure reporting and feedback, co-ordination of failure data, preparation of operating and maintenance instructions, failure and stress analysis leading to reliability prediction and to design trade-offs, contract control, prototype testing, control over manufacture. Some of these activities are also required for a programme of maintainability design and will be examined in due course.

Maintainability is achieved by a combination of maintenance philosophy and product design. Location of spares and equipment is an example of the former, whereas the allocation of test points and built-in test gear are examples of product design. Methods of achieving maintainability are dealt with in the next two chapters.

The activities which are necessary to ensure that system design and choice of maintenance methods result in optimum maintainability form a maintainability programme. Since these activities overlap with those for achieving reliability, and since reliability and maintainability often interact in determining availability, a single integrated reliability and maintainability programme is usually desirable. The following list shows the activities which must be scheduled into the design, manufacturing and commissioning programme and indicates their relevance to maintainability (M) and/or reliability (R).

Decide Objectives, Responsibility and Plans (M, R)
Objectives may be fixed by contract; otherwise they must be decided by a consideration of the requirements and environment of the system use. Responsibilities should be clear from company system effectiveness policy and a programme plan prepared by the project design engineer or manager responsible for reliability and maintainability.

Train Personnel (M, R)
Design engineers should be trained to a level where they can work with the system effectiveness expert rather than the expert having to demand design changes.

Contract Control (M, R)
Statements of failure rate and probability, mean time to fail and repair time must be accompanied by definitions of failure, environment, maintenance conditions, extent of liability, method of demonstration and other important factors. Such a contract should not be accepted without reference to the system

effectiveness expert. If work is to be sub-contracted then in order to meet objectives it may be necessary to make contractual demands on sub-contractors in terms of maintainability and reliability parameters.

Stress and Failure Analysis (M, R)
A study is necessary of the environmental and circuit stresses applicable to each component together with a study of the effect on the system should that component fail.* This enables a reliability prediction to be made, and at the same time and as part of the maintainability analysis, provides an opportunity to check that each failure is both diagnosable and easily repairable.

Reliability Prediction (R)
Requires failure rate data, stress and failure analysis and statistical techniques.

Maintainability Analysis (M)
This involves the analysis of the maintenance requirements dictated by the proposed design and the trade-off of these requirements against design alternatives. Factors include: built-in test equipment, external test equipment, least replaceable assembly (the level at which diagnosis ceases and a replacement is made—for example a printed card), throw-away level (a function of cost and failure rate), support equipment, manuals, skill and training requirements.

The repair time, dictated by these and design considerations, can be predicted in order to compare alternative combinations of maintainability factors. Some combinations may be less expensive than others. The following chapters examine these factors and some methods of repair time prediction.

The maintainability analysis must also include an appraisal of the preventive maintenance requirements of a particular design. As with corrective maintenance, service times can be predicted and the likely costs forecast. Some maintenance facilities or equipment may be fixed by reason of the customer's operating environment, and the design must then be tailored to accommodate them. Access and test points are design features which may have to be brought into line with these maintenance conditions.

Design Review (M, R)
Reliability and maintainability should be monitored throughout design, changes and trade-offs being made at the earliest possible point. A design review body, besides ensuring that these activities are being carried out, can monitor the progress of design towards the system effectiveness goals. The design review team should consist of representatives of design, marketing, purchasing and manufacturing, but should never be chaired by someone involved in the design of the product. Several design reviews should take place

* This is known as the *failure mode and effect analysis* and is often abbreviated to FMEA.

at intervals to be decided in the light of the progress of the design and the nature of the problems encountered.

Design Trade-offs (M, R)
These will take place between methods of achieving reliability and maintainability and may even involve sacrificing one for the other as, for example, in forgoing the reliability of a wrapped joint in favour of the maintainability advantages of a connector. Major trade-off decisions may well be taken at design review meetings, whereas less important trade-offs may be made by the system designer and system effectiveness engineer.

Allocation (M, R)
Maintainability and reliability objectives may well be set for the whole system, and it then becomes necessary to allocate objectives to each separate unit of equipment. Since design effort is often limited it is desirable to allocate more cost and effort to the repairability of high failure rate situations than to those which occur less frequently. Similarly, maximum reliability effort may be given to the least reliable units.

Cost Recording (M, R)
In the evaluation of the effectiveness of the maintainability and reliability effort it is useful to know how much has been spent to this end. A difficulty exists, however, in deciding which costs would have been incurred even with no specific maintainability or reliability objectives.

Data Collection and Co-ordination (M, R)
Accurate and detailed failure reporting is essential if maintainability design is to be effective and if actual maintenance experience is to be fed back into the design of the system in order to improve effectiveness. Failure data together with information concerning the stress conditions applicable is also essential to reliability prediction.

Analysis of Failures (M, R)
Permits lessons to be learnt and fed back to design.

Prototype Testing (M, R)
Provides an opportunity to test both maintainability and reliability and to compare the results with previous theoretical assessments.

Maintainability Prediction (M)
Can be obtained by examining a production model or from detailed design information. Methods are outlined in a later chapter.

Manufacturing Controls (R)
Ensure design tolerances are adhered to; otherwise the system becomes less tolerant to changes in environmental stress. This includes the issue of manufacturing and quality specifications.

Documentation (M, R)
Operating instructions and maintenance manuals play an important part in assuring reliability and a vital part in achieving maintainability objectives.

Part Selection and Control (R)
Involves supplier selection and contracts. Continued and alternative supplies are a consideration.

Spares Provisioning (M, R)
Involves customer liaison. Affected by reliability.

Burn-in (*Pre-stressing*) (M, R)
Usually performed to improve reliability by removing early failures.

Demonstration of Maintainability (M, R)
As with reliability, maintainability can only be demonstrated on a sample basis and therefore a statistical risk is involved for both parties. This is also dealt with in a later chapter.

Clearly the activities listed above will assume a relative importance to each other and to other design activities according to the product type, size of company, attitude of customer, and so on. The order in which they are carried out will not always be the same and will, almost certainly, not be the order in which they have been presented here. Furthermore, they will overlap with each other and the task of preparing a programme plan will involve scheduling them along with other design activities upon which these maintainability and reliability activities are dependent. Maintainability analysis, for example, cannot commence until both circuit and equipment design is well advanced, whereas an initial reliability prediction requires only tentative circuit details and system configuration.

One formal guide to the preparation of a maintainability programme is the US Military Standard 470, *Maintainability Program Requirements for Systems and Equipments*.

It has already been stated that mean time to repair (MTTR) is the parameter of maintainability, and the reader will recognize the latter as a statement of the probability of achieving some value of the former. The relationship between the two is hence determined by the way in which the statistical variable, MTTR, is distributed. In the same way reliability (probability of non-failure) is determined by the distribution of times to failure. Repair times can be subdivided into several parts such as diagnosis time, replacement time, transportation of spares time, etc., and the times for these activities are not distributed identically. These activities can be grouped under the two broad headings of *active* and *passive* repair time. These will be dealt with in the next chapter, but broadly speaking, active repair time covers diagnosis and repair of the system (directly influenced by design), whereas passive repair time covers administrative and logistic times (not influenced by design). It has been observed that active repair times are usually distributed on a log normal basis, and that administrative times follow a Weibull distribution (i.e. maintainability is of the form $1 - \exp(-t^{\beta}/\alpha)$. A consideration of the distribution of repair times is necessary if MTTR is to be expressed as maintainability or if a statistical statement is to be made concerning MTTR obtained from sample measurements of the repair times for certain faults. For other purposes it is often sufficient to deal with the parameter MTTR for the subdivision of repair time in question. For the purpose of this book, then, it will only be necessary to consider distribution of repair times in the chapter devoted to the statistical demonstration of maintainability.

The cost of achieving a maintainability objective consists of the costs of design, manufacture, test equipment, manuals, spares, transport, etc., and a maintainability objective can be achieved by a combination of these factors. Trade-offs exist between such factors as quality and quantity of test equipment and the use of more skilled manpower, between mechanical design and least replaceable assembly (level at which replacement takes over from diagnosis), between training of personnel and detail of maintenance instructions, etc. The most economical combination of factors resulting in the required objective is dictated by the environment, the system user and his maintenance organization, the availability required from the system and other factors peculiar to the system and the customer. The above subdivision of costs will be seen to fall into the two broad categories of costs to the manufacturer and costs to the customer. Some of these, such as the cost of test equipment and its maintenance and the cost of replacement parts, can fall to either party according to the contractual terms prevailing. Improved maintainability may increase the price of a product, but, as with reliability, will decrease the operating costs. The minimum cost of ownership (capital cost and operating cost) occurs for some particular value of reliability and maintainability.

Fig. 1.1 shows the general relationship between availability and price. The curve of cost of design and manufacture (costs before delivery) increases with availability, and the curve of costs of contractual repair, poor publicity, etc. (costs after delivery) decreases with availability. The sum of these curves yields a curve which, being the total costs of the supplier, must relate to price.

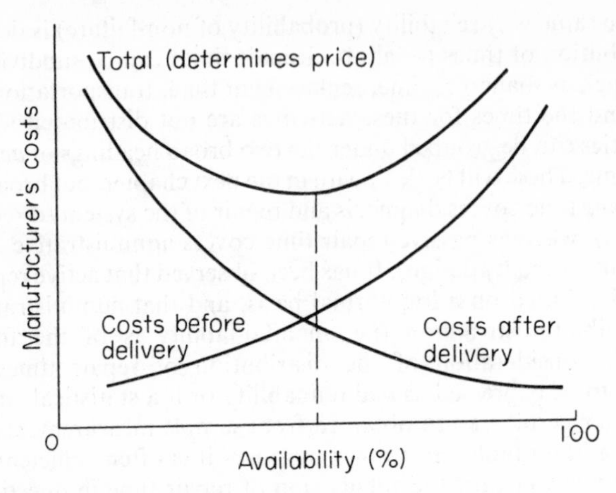

Fig. 1.1 Price and availability

As can be seen from the diagram there is a value of availability which corresponds to minimum price.

Fig. 1.2 shows how price and the operating and maintenance costs of the customer combine to yield the curve of total cost of ownership against availability. Again there is a value of availability corresponding to minimum cost, but in this case it occurs at a higher availability than for the supplier's minimum cost.

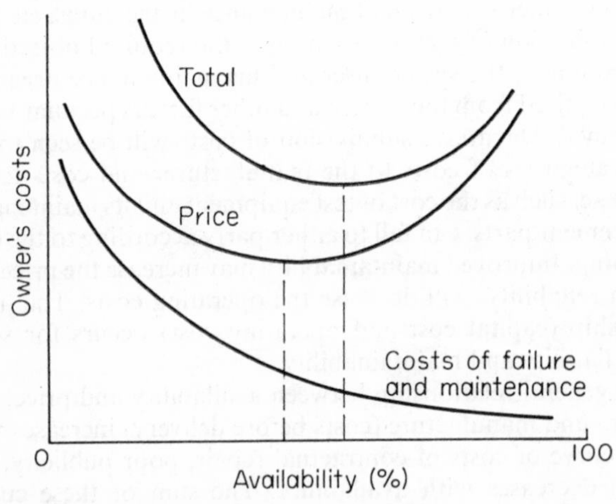

Fig. 1.2 Cost of ownership and availability

The cost of providing reliability and maintainability parameters, the relationship between them and the actual cost of maintenance are a complex interaction about which it is impossible to make general statements. This state of affairs is by no means peculiar to system effectiveness since other system parameters such as power consumption, operating personnel costs, revenue from output in different modes of system operation interact in a complex fashion. Figs. 1.1 and 1.2 consider the apportionment of costs between consumer and producer and how these vary, as a whole, with availability; and Fig. 1.3 illustrates that these parameters and costs are mutually interdependent.

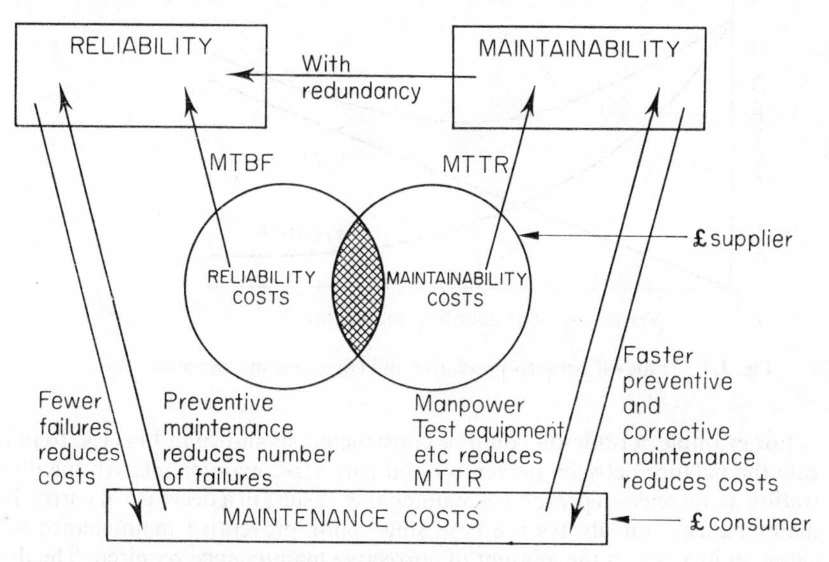

Fig. 1.3 Interrelationship of system effectiveness and costs

Money spent on maintainability reduces repair time, which in turn improves reliability in the case of failed redundant units. Improved reliability reduces maintenance costs, whereas money spent on preventive maintenance may enhance reliability. A trade-off situation exists with maintainability and maintenance: money spent on manpower and test equipment improves maintainability by reducing maintenance time; whereas money spent on accessibility, test points, built-in test gear and displays reduces both preventive and corrective maintenance time and hence cost.

Where it is required to make a trade-off involving one or two parameters and few alternative solutions exist then a relatively simple analysis of benefits can be made in terms of cost or repair time or mean time to failure. Clearly it is not possible to optimize every effectiveness parameter and every cost benefit, and some combination must be found which offers an overall optimum benefit in terms of system effectiveness and cost. It is not possible to consider, simultaneously, every system variable in order to arrive at a combination which identifies the optimum points on the curves of Fig. 1.1 and 1.2. It is necessary to compare costs of different methods of achieving each variable, where alternatives exist, and to consider the advantages and penalties of sacrificing one parameter for another.

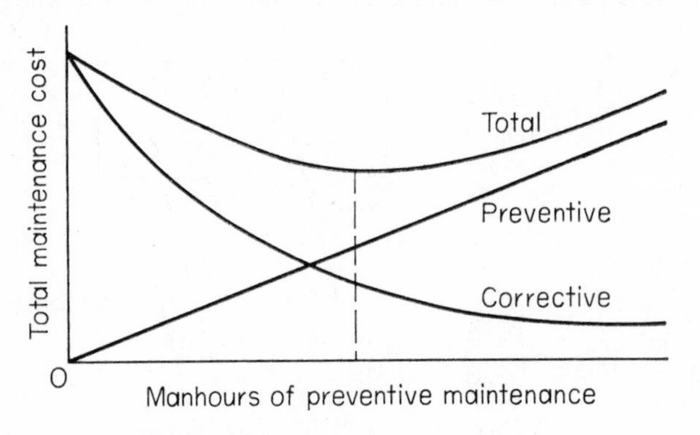

Fig. 1.4 Trade-off between preventive and corrective maintenance

For example, a trade-off could be constructed, as shown in Fig. 1.4, to indicate the balance between preventive and corrective maintenance. In the illustration an increase in preventive maintenance results in a decrease in corrective maintenance—not always the case since poor preventive maintenance will cause an increase in the amount of corrective maintenance required. The diagram illustrates the existence of an optimum combination of these two factors.

Preventive maintenance activities are of four types. The first is the *regular care* type of maintenance which involves cleaning, oiling and adjusting. The maintenance actions actually carried out and their frequency are determined by the nature of the equipment and its operating environment.

The second type of preventive maintenance is known as *preventive replacement* and applies to the increasing failure rate situation associated with wearout. If the statistical distribution of failures is known then the probability of failure up to any suggested replacement time can be assessed. If a suitable risk of failure due to wearout is chosen then the time at which replacement should take place can be calculated. Where the cost of access to a system is high all items of a particular type may be replaced (group replacement) during the preventive maintenance action.

The third category is that of *identification of dormant faults*. Many faults do not cause immediate system or unit failure and give rise to no alarm condition. A further fault or some unusual combination of stress conditions is often required before the first fault becomes apparent. An example of a dormant fault is an alarm which, having itself failed, cannot indicate a system failure. This is a particularly dangerous situation since the effect is to conceal a failure. Any checking procedure which results in the identification of dormant faults will improve reliability.

The fourth type of preventive maintenance is the *identification of degradation failures*. Regular measurements, aimed at identifying parameters which are drifting towards the limit of acceptable performance, could result in preventive replacement of the offending component before a failure occurred. This type of maintenance requires a great deal of routine testing and is therefore expensive in terms of manpower and test equipment. Additional test points have also to be provided in order to permit this type of observation. A serious drawback is the danger of causing failures due to the disturbance occasioned by the maintenance technician.

The absolute cost of providing maintainability and reliability is impossible to assess since every aspect of design and manufacture can add to or detract from these parameters. It is meaningful, however, to consider the cost of that design effort, documentation, training, etc., provided specifically to achieve given maintainability and reliability objectives. Such an assessment is a measure of the additional cost to provide the increment of system effectiveness above that which would exist if no particular effort had been made.

Summary

Quality is adherence to specification: failure is departure from specification.

$$\text{Cost of failure} = \text{Cost of repair} + \text{Cost of outage (revenue)}$$

Total cost of failure is determined by reliability and maintainability (failure rate and repair time) and specific objectives are obtained by:

Stating objectives; training; contract control; stress and failure analysis; reliability and maintainability prediction; maintainability analysis; design reviews; design trade-offs; data collection and feedback; testing; manufacturing controls; maintenance manuals; analysis of failures; supplier and parts selection and control; spares provisioning; burn-in; demonstrations.

Repair times are distributed about a mean value. The distribution is sometimes log normal, sometimes Weibull, sometimes bimodal and sometimes unknown.

There is a minimum cost of ownership value of availability (Figs. 1.1 and 1.2). Fig. 1.3 shows the interaction between reliability, maintainability and their costs.

Preventive maintenance consists of routine care, preventive replacement, identification of dormant faults and the identification of degradation failures.

Individual trade-offs may be computed as in Fig. 1.4.

The cost of providing system effectiveness improvements can only be expressed in terms of increments.

2

Down Time and Repair Time

The *down time*, or outage, of an equipment is the period during which it is not in an acceptable working condition; however, a formal definition is usually avoided due to the difficulties of generalizing about a parameter which may consist of different time elements according to the system and its operating conditions. Consider the following examples which help to emphasize the difficulty:

(*a*) A system not in continuous use may develop a fault at a time when it is idle. The fault condition may be immediately apparent to the user or it may not become evident until the system is required for operation. Is down time to be measured from the incidence of the fault, from the start of an alarm condition or from the time when the system would have been required?
(*b*) In some cases it may be economical or essential to leave an equipment or even a system in a faulty condition until a particular moment or until several similar failures have accrued.
(*c*) A repair may have been completed but it may not be safe or convenient to restore the system to its operating condition immediately. Alternatively due to a cyclic type of situation it may be necessary to delay. When does down time cease under these circumstances?

It is necessary, as can be seen from the above, to define down time as required for a particular system under given operating conditions and maintenance arrangements. A further confusion arises when distinguishing between repair time and down time. The two, although overlapping, are not identical as in the case of a delay between actual loss of an operating system and the realization of that fact by the system user. Down time, in this case, commences some time before any repair activity. Repair usually involves some checkout and realignment elements and these might extend beyond system outage. Here the down time finishes before the repair activity. Our definition and use of these terms will depend on whether system availability or maintenance manhours are under consideration. Again, the terms have differing significance at the system level, the level of a redundant unit or the replaceable module level. If the system function is restored by the insertion of a replaceable module then the system down time and repair time are not influenced by the time to repair the failed module although that time influences manpower costs and the availability of spares.

In general a change in system repair time results in a change of down time but, as has been pointed out, the two are not entirely interdependent.

The following table identifies the effect of down time and repair time on both system and subsystem levels. Note that, although overlapping, down time and repair time are not identical. This will be further discussed in relation to Fig. 2.1 (page 20).

	Down time	*Repair time*
System	Determines availability and hence cost of lost revenue	Contributes to cost of maintenance*
Redundant unit	Influences system reliability	Contributes to cost of maintenance*
Replaceable module	Influences spares availability	Influences both cost of maintenance and availability of module as a spare part

* The cost of maintenance is mainly determined by the manhours required to effect a repair (excluding preventive maintenance at present), but the quantity of manhours is not exactly proportional to the repair time owing to the existence of periods of inactivity due to administrative and logistic delays.

It is down time which determines availability, and this may therefore be expressed as follows:

$$\text{Steady-state availability} = \frac{\text{Up time}}{\text{Up time} + \text{Down time}}$$

$$= \frac{1}{1 + \lambda \,.\, \text{MTTR}}$$

$$= \frac{\text{MTBF}}{\text{MTBF} + \text{MTTR}}$$

Up time + down time is equal to the total time for which the system is required to function. This may be unbroken real time (e.g. telephone exchange) or it may be a proportion thereof (e.g. mobile intercom.); λ is the failure rate of the system; MTBF, the mean time between failures; and MTTR, the mean time to repair of the system. The expression in terms of MTBF and MTTR assumes that MTBF $= 1/\lambda$; this is only true when constant failure rate applies.

It has already been stated that availability is determined by down time and not by repair time; however, the term MTTR is widely used to describe both.

Care must be taken to define what elements of system outage and or repair have been used in the statement of a value of MTTR.

Instantaneous availability is the probability of being capable of operation at some random point in time: steady-state availability can be shown to be the integral of instantaneous availability from zero to infinity. Down time and reliability (expressed above in terms of failure rate) determine availability, but in addition down time can influence reliability when redundancy applies. The time to repair a redundant unit must influence the probability of system failure due to a further fault occurring during repair of the first unit. It can be shown, for example, that if two units are in a state of active redundancy and the failure rate and MTTR of each are λ and $1/\mu$ respectively then for that system

$$\text{MTBF} = \frac{3\lambda + \mu}{2\lambda^2}$$

Once again the MTTR refers to the mean outage of a unit. An explanation of the derivation of the above can be obtained by referring to a textbook on reliability.

System down time may be divided into a number of elements, of which the following are the main ones.

(*a*) *Time to Realization*

The time which elapses before the fault condition becomes apparent may depend on whether or not the system is attended, on the alarm arrangements, on the type of service the system gives (as to how readily non-operation would be noticed). This element of down time is pertinent to availability but does not constitute part of the repair time.

(*b*) *Access Time*

This involves the time, from realization that a fault exists, for the maintenance man or team to make contact with the displays and test points of the system and so commence fault finding. This does not include travel of men and equipment but the removal of covers and shields and, where necessary, the connecting in of test equipment. This is determined by mechanical design considerations.

(*c*) *Diagnosis Time*

This is often referred to as fault finding or "faulting." It includes the warming up and adjustment of test equipment (e.g. warming up and setting the controls of an oscilloscope or signal generator), the carrying out of checks (e.g. examination of waveforms or voltages for comparison with those given in a handbook), interpretation of the information gained from test equipment (this may be aided by algorithms or other maintenance instructions), verifying the conclusions drawn from the above and deciding what corrective maintenance action to take. This element is influenced by the level at which fault identification is to take place (component, card, shelf, etc.), by the tools, equipment, maintenance instructions and skill of personnel, by the location of test points and built-in test equipment, by the displays, and by the number of maintenance men assigned to the task.

(d) Spare Part Procurement Time

Part procurement can be from the tool box, by cannibalization or by taking a redundant identical assembly from some other part of the system. The latter is only resorted to if all else fails since the removal of a "good" unit from the system is highly undesirable. Some spares may be kept on site and in some cases spare units may be located in dummy positions alongside active units. Procurement by any of the above means is included in this repair category, but the time taken to move parts from some depot or central store due to their not being available on site is not included. This activity is referred to under logistic time (*h*).

(e) Replacement Time

This involves access to and removal of the faulty least replaceable (LRA) assembly, installation and connection of the new LRA, rewiring and repairs to wiring if plug-in connectors are not used, and resecuring and cleaning as required. The LRA is the replaceable item below which fault diagnosis does not continue. Replacement time is largely dependent on the choice of LRA and on mechanical design features such as the securing of the LRA and the choice of connectors.

(f) Checkout Time

This involves verification that the faulty condition has ceased and that the system is operational to the appropriate standard. This activity should be aided by precise and unambiguous instructions in the maintenance manual. As with diagnosis times, it is largely influenced by the degree of instructions given and by the type of test equipment used. A warm-up period may be necessary first. It may be possible to restore the system to operation before completing checkout, in which case, although a repair activity, it does not all form part of the down time.

(g) System Alignment Time

The association of a new LRA with the system may require the adjustment of both in order to optimize, or even provide, performance. As with checkout time, some or all of the alignment time may fall outside the down time.

(h) Logistic Time
This is the time spent waiting for spares to be transported to the site and for the test gear, additional tools and manpower to be brought to the system. The greater the number of non-standard parts and of different LRAs in the system, the greater will be the probability of lengthy logistic delays, and in this sense logistics is determined, to some extent, by design considerations. It is also governed by manpower and maintenance considerations and by spares provisioning.

(i) Administrative Time
This is a function of the system user's organization. Typical elements involve failure reporting (where this affects the down time), allocation of repair tasks, manpower changeover due to demarcation arrangements, official breaks, disputes, etc.

Activities (*b*)–(*g*) are called *active repair times*, and activities (*h*) and (*i*), *passive repair times*. Realization time is not really a repair activity but may be included in the MTTR where down time is the consideration. Checkout and alignment, although utilizing maintenance manpower, can fall outside the down time of the system. The active repair activities are determined by the design of the equipment (i.e. inherent maintainability) and by the maintenance arrangements, environment, quantity and skill of manpower, maintenance instructions and by the tools and test equipment. Logistic and administrative time, on the other hand, is mainly determined by the maintenance environment rather than by equipment design. In other words the location of spares, equipment and manpower, and the procedures for allocating tasks and for failure reporting during the repair, influence passive times but are not under the control of the designer. Design will, however, have some influence on these, as has been pointed out in the case of component standardization minimizing logistic delay.

The active elements of repair time will often occur in the order in which they have been presented, whereas the passive elements as well as elements of access time will occur at any time during or between the times of other activities. Fig. 2.1 refers to the foregoing.

There is a degree of interdependence between these active and passive activities. As the active repair time increases there is a greater incidence of human rest periods, logistic delays for the provision of spares and additional tools and test equipment and of administrative delays interfering with the repair. Recycling of activities can occur as in the case of incorrect diagnosis or the case of a replacement LRA being found to be faulty. Another cause of recycling would be the incidence of a fault occasioned by the repair activity itself. Every effort must be made to ensure that incorrect diagnosis will not occur since the removal of a replaceable module which is not faulty is highly undesirable both from the point of view of the danger of inducing further faults and from the point of view of the down time itself.

Since maintainability is defined in terms of the probability of not exceeding a stated repair time (or down time) it is therefore dependent on the way in which these times are distributed. MTTR (Mean time to repair) will be used as the parameter of maintainability, and it will be assumed that the appropriate elements are present in that MTTR according to the circumstances. A system

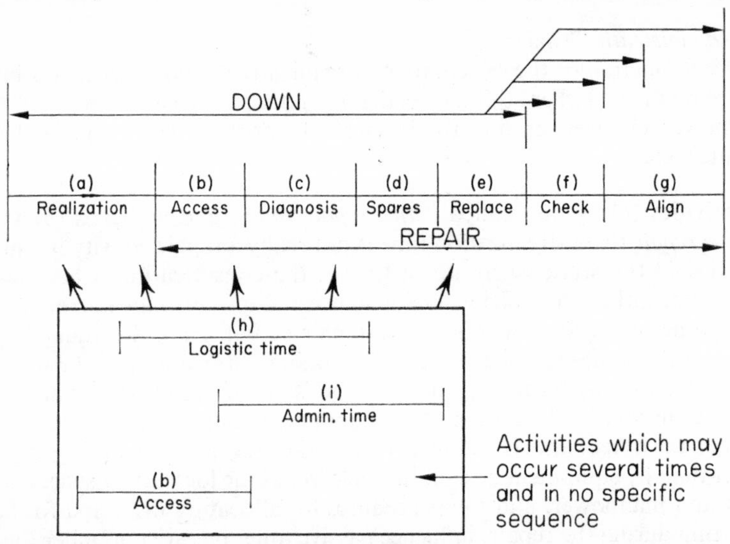

Fig. 2.1 Elements of down time and repair time

will have as many failure rates as there are modes of failure applicable to it, and in the same way there will be a similar number of repair times each expressible as a mean together with some distribution. It is also worthwhile to consider the average repair time for all types of fault, provided that the individual MTTRs in the calculation are weighted according to their relative frequencies of occurrence (determined by the failure rate for each failure mode). It would then be necessary to know how this overall average is distributed in order to express it in terms of a maintainability.

Consider that there are x failure modes of a system each characterized by a failure rate λ_i. Let there be y repair actions observed for each mode each having repair time $1/\mu_j$. Each individual MTTR_j for each failure mode is therefore

$$\sum_{j=1}^{j=y} \frac{1/\mu_{ij}}{y}$$

The overall weighted MTTR is then

$$\frac{\sum_{i=1}^{i=x}\left(\lambda_i \sum_{j=1}^{j=y} \frac{1/\mu_{ij}}{y}\right)}{\sum_{i=1}^{i=x} \lambda_i}$$

If the mathematical model of the statistical distribution of some quantity is known then it is possible to state a probability for any value of that quantity to fall within given limits if drawn at random from the population in question. In the case of a normally distributed quantity, for example, the probability of an item taking a value between plus or minus one standard deviation of the mean is approximately 68%.

In the case of maintenance activities, active repair times are usually said to be distributed according to the log normal rule; in other words, the logarithms of the times are normally distributed. This may be expressed as follows:

$$M(t) = \int_0^t \frac{1}{t\sigma\sqrt{(2\pi)}} \exp\left[-\tfrac{1}{2}\left(\log_e \frac{t-d}{\sigma}\right)^2\right] dt$$

where $M(t)$ = Probability of a repair time not exceeding t, i.e. the maintainability

t = A possible repair time

σ = Standard deviation of $\log_e t$ about the mean of $\log_e t$

d = Mean of the values of $\log_e t$

Passive repair activities are often described by means of the Weibull rule, in which case

$$M(t) = \int_0^t \left[1 - \exp - \frac{t^\beta}{\alpha}\right] dt$$

where β and α are parameters of the distribution which can most easily be found by a graphical analysis of specimen repair times for the relevant activities.

In both the above cases, if t is much greater than the mean of t (i.e. MTTR) then the following approximation is possible:

$$M(t) \approx 1 - \exp - t/\text{MTTR}$$

As with many statistical quantities it is often advisable to construct a histogram of the data in order to illustrate the outline of the distribution. This has the advantage of showing up obvious departures from the expected distribution. More sophisticated techniques exist for checking that a distribution conforms to a given mathematical model, but these will not be dealt with here.

Fig. 2.2 shows a typical log normal distribution and a bimodal distribution whose mathematical model we shall not concern ourselves with.

Such a bimodal distribution of repair times would indicate that, in fact, two distributions were present in the data. In other words two repair actions could be included in one set of repair times and this would yield a histogram catering for two means. Other possibilities include sets of data referring to two sets of maintenance personnel, two sets of maintenance instructions, different environments, etc.

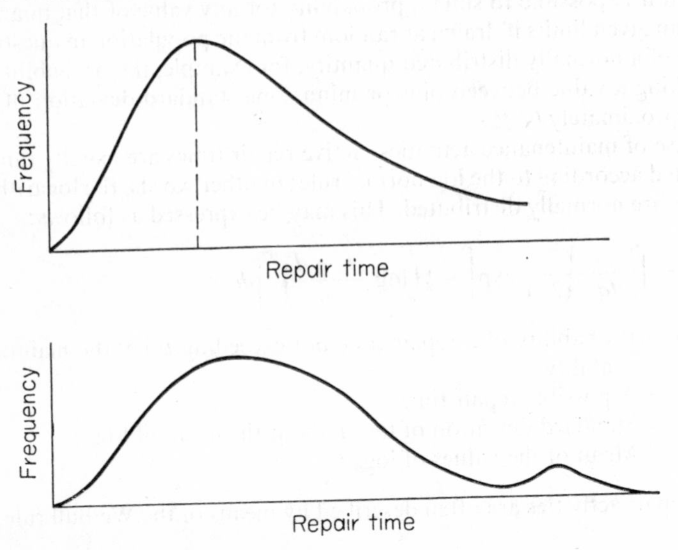

Fig. 2.2 Repair time distributions

In the next chapter we shall be concerned with methods of reducing active and passive repair times and hence improving maintainability and reducing both preventive and corrective maintenance costs.

Data concerning the various elements of down time can be obtained either from tests carried out on a prototype or production model of from the field. In either case some formal failure reporting document is necessary in order to ensure that data are both consistent and adequate in detail. Field data are by far the more valuable of the two types since they refer to failures and maintenance actions which have taken place under real conditions. Field data are, however, expensive to collect and co-ordinate, and since their acquisition depends on persons rather than equipment, they are subject to errors, omissions and misinterpretations. Information of this type has two uses. The first is feedback for the modification of both equipment design and repair procedures. The second is the acquisition of statistical reliability and maintainability data. Summarizing the reasons for collecting failure data then:

They indicate design and manufacturing deficiencies
They indicate quality and reliability trends
They provide sub-contractor ratings
They contribute to statistical data for future predictions of reliability and maintainability.
They assist second-line maintenance.

The collection of failure data requires appropriate documentation and fully trained and enthusiastic personnel.

A failure reporting system should be established for every project or product and should collect failure data from prototype and production models and, wherever possible, from field systems. Customer co-operation with a failure reporting system is essential if feedback from the field is required and this may well be sought, at the contract stage, in return for some other reliability or maintainability concession demanded from the supplier.

The failure reporting form must record information on the following:

Repair time: active and passive.
Type of fault: i.e. primary or secondary, random or induced by maintenance action or by unusual environment.
Nature of fault: open-circuited or short-circuited component, drift condition, etc.
Location of fault: exact position of LRA or component concerned. Geographical location of system.
Environmental conditions: if possible the environmental conditions at the time of the fault should be recorded. This is not always possible since the fault may have occurred due to or during earlier stress conditions.
Action taken: exact nature of repair action. Name of item repaired or replaced.
Maintenance personnel: numbers of and names of personnel carrying out repair. Enables information to be gathered about individual maintenance personnel, training to be evaluated, etc.
Equipment used: test equipment and tools used to effect the repair. Enables usage of equipment to be monitored.
Spares used: allows spares provisioning to be checked, together with costs of repair and state of spares stocks.
Build specification: changes to manufacturers' specifications may result in the use of different processes, materials, standards or components, and these changes may have a relevance to the failure condition. This information may be recorded by quoting a serial number or drawing reference.

ITTE Failure Report and Action Form

	System	Sub-system	Module/ sub-assembly
Type			
Serial number			
Location/identification			
On-time (cumulative)			
Down time (this failure)			
Active repair time			

Report number:
Report date:
Report completed by:
Company:

To be completed on site

System status:
- Field service ☐
- Field trial ☐
- Production prototype ☐
- Model ☐

Details of symptoms, diagnosis and failure:

Effect of failure on system:
- Complete system failure ☐
- Major degradation ☐
- Minor degradation ☐
- None ☐

On-site diagnosis:
- No defect found ☐
- Part failure ☐
- Installation defect ☐
- Manufacturing defect ☐
- Design defect ☐
- Program defect ☐
- On-site human error ☐
- Other ☐

Action taken:
- Replace module ☐
- Repair ☐
- Modification ☐
- Program reload ☐
- Other ☐

Details of action taken:

To be completed at Designated Centre

Project engineering action:		Name	Company/Dept	Signature	Date completed
– Consolidate with filed data	☐				
– For immediate analysis/action by:					
Engineering	☐				
Manufacturing	☐				
Quality assurance	☐				
Purchasing	☐				
Other	☐				

Analysis and action taken:

Engineering change no.:
Dated:
Follow-up report. Ref no.:
Dated:

Name:
Signature:
Date:

For information to:

Fig. 2.3 ITT Europe failure report and action form

The main problems associated with failure recording are

Motivation: if the maintenance technician can see no purpose in recording information in the detail of Fig. 2.3, it is likely that items will be either omitted or incorrectly recorded. The purpose of fault reporting and the ways in which it can ultimately be used to simplify his task should be explained to the technician during training. If he is frustrated by unrealistic time standards, poor working conditions, inadequate instructions, etc., then the failure report is the first task which he will complete hurriedly and inaccurately.

Verification of accuracy: once the failure report has left the person who completes it the possibility of subsequently checking its accuracy is very much reduced. If repair times are suspect or if incorrect diagnosis of failure causes is suspected then it is likely that these inaccuracies will go undetected or, at least, unverified. Where failure data are obtained from customer's staff in the field the possibility of challenging information becomes even more remote.

Cost: failure reporting is costly both in terms of the manhours used to complete failure report forms and in terms of the manhours of co-ordination and interpretation of the information obtained. For this reason both supplier and customer managements are often reluctant to agree to a comprehensive failure reporting system. If the information is correctly interpreted and action taken within design or manufacturing to remove sources of failure then the cost of this activity is likely to be offset by the savings in failure costs.

Recording of non-failures: a particularly undesirable reporting situation occurs when a failure is recorded where no failure exists. This occurs as a result of two types of mistake on the part of the technician. Firstly he is likely to test for a fault by replacing suspect but not necessarily failed components. When the fault eventually disappears he does not then remove the first, wrongly replaced components and these may be recorded as failures. Failure rate data are hence artificially inflated. The second problem concerns the interpretation by the technician of secondary failures as primary failures. A failing component may cause stress conditions to apply to another component which result in its immediate failure. Diagnosis will reveal both failures but not necessarily which one occurred first. Again failure rate data become artificially inflated. More complex maintenance instructions and the use of higher-grade better-trained maintenance personnel will help to reduce these problems but may increase costs beyond a reasonable level and will certainly never totally remove incidents of incorrect failure recording.

Fig. 2.3 shows an example of a well-designed and thorough failure recording form as used by the European Companies of the International Telegraph and Telephone Corporation. This single form strikes a balance between the need for detailed failure information and the requirement for a simple reporting format. A feature of the ITT form is the use of four identical print-through forms. The information is therefore accurately recorded four times with minimum effort.

As an example of the need for more elaborate reporting, consider an air traffic control service which is essentially concerned with the safety of life in a dynamic situation. The actual security and availability of all its systems must be known. It is not surprising, therefore, to find an appropriately detailed maintenance reporting system in such an organization. Three of the forms used by the British Civil Aviation Authority are reproduced in Figs. 2.4, 2.5 and 2.6. These forms deal with preventive and corrective maintenance reporting on the operational site, and also the reporting of workshop repairs of "off-line" items. Earlier versions of these forms were designed with the requirements of computers used for routine analysis in mind. Redesign successfully combined these requirements with that of ready comprehension by a human reader.

It is unfortunate for the designer, seeking to fulfil maintainability objectives, that few reporting forms give adequate breakdown of maintenance times with separation of passive and active elements. To identify and record this information adds to the maintenance load and hence to the cost. This cost must be justified if a special investigation is to be set up. Such an investigation, continued as long as there is useful development of maintenance procedures or of equipment design for maintainability, will pay for itself by reducing long-term maintenance costs.

Summary

Down time is not the same as repair time.

Availability = MTBF/(MTBF + MTTR), where MTTR refers to the down time.

Down time is made up of realization, access, diagnosis, procurement, replacement, checkout, alignment, logistics and administration.

Maintainability function is determined by the distribution of repair times.

Active elements are assumed to be log normally distributed.

Passive elements are assumed to be Weibull distributed.

Beware of bimodal distributions.

Failure and event recording provides vital information on design, subcontractors and manufacturing control as well as providing statistical information to assist reliability predictions and second-line maintenance.

The main problems associated with maintenance recording are: motivation, verification, cost, recording of non-failures, and inadequate breakdown of maintenance times.

CA 1510 REVISED APRIL 1970	NATIONAL AIR TRAFFIC CONTROL SERVICE	CORRECTIVE MAINTENANCE REPORT

A GENERAL SECTION

	FOR USE OF ADP STAFF ONLY	TYPE OF REPORT (CIRCLE ONE ONLY)	LOCATION	REPORT SERIAL NUMBER
(N	&Ø1	&Ø2 C P I F	&Ø3	&Ø4

&Ø5 CAUSE OF INCIDENT (CIRCLE ONE ONLY)

Damage Due To Weather	Ø1	Service Failure Due To Weather	11	O.E.S. and St/By Failure	21	Software Fault	31	Aircraft Fault	41
Damage Due To Vandalism etc.	Ø2	Propagation	12	O.E.S. Failure St/By Running	22	Suspected Software Fault	32	Maintenance Activity	42
Electronic Fault	Ø3	Interference	13	Works Services	23	Suspected Hardware Fault	33	Cause Unknown	43
Mechanical Fault	Ø4	Weather Interference	14	Mains Distribution	24		34	Multiple Trips	44
Private Wire Fault	Ø5	Site Infringement	15		25		35	No Fault Found	45
	Ø6		16		26		36	Other Reason	46

B SERVICE INTERRUPTION SECTION

FACILITY	DATE TIME OF SERVICE INTERRUPTION (Day Month Time)	SERVICE DOWN TIME (Hrs. Mins.)	
&Ø6	&Ø7	&Ø8	&Ø9

How was interruption detected? (Circle One Only)

Aircraft Report	A	Manual Monitor	D
Other User Report	B		E
Automatic Monitor	C		F

C EQUIPMENT DEFECT SECTION

DATE TIME OF INCIDENT (Day Month Time)	DURATION OF INCIDENT (Hrs. Mins.)	WORKING TIME IN MAN HOURS (Hrs. Mins.)	TRAVELLING TIME IN MANHOURS (Hrs. Mins.)	DETECTED BY MONITOR SYSTEM?	EQUIPMENT
&1Ø	&11	&12	&13	&14 Y N	&15

DESIGNATION	ASSEMBLY	SUB ASSEMBLY	ITEM	ACTION	
&16	&	&	&	&	& &
&17	&	&	&	&	& &
&18	&	&	&	&	& &
&19	&	&	&	&	& &
&2Ø	&	&	&	&	& &
&21	&	&	&	&	& &

ACTION CODES

Replaced, Open Circuit	A
Replaced, Short Circuit	B
Replaced, Intermittent	C
Replaced, Over Heated	D
Replaced, Out of Tolerance	E
Replaced, Other Reason	F
Adjusted	G
Repaired	H
Wiring Fault Repaired	J
Cleaned	K
Item Dried Out	L
Lubricated	M
Rubber Moulding Perished	N
Refurbished	X
Sent to Industry	Y
Scrapped	Z

&22 D SPECIAL INVESTIGATION SECTION

&23	&24	&25	&26	&27	
&28	&29	&3Ø	&31	&32	
&33	&34	&35	&36	&37	&?

E COMMENTS

Name and Grade in Block Capitals ....

DISTRIBUTION:- Tels HQ – Green STO – Blue
Division – Red Equipment Records – Black

CHARACTERS	
Figure One	1
Figure Nought	Ø
Letter I	I
Letter O	O

DATE TIME INCIDENT CLOSED

CROSS REFERENCE

Fig. 2.4

CA 1511 REVISED APRIL 1970	NATIONAL AIR TRAFFIC CONTROL SERVICE	PLANNED MAINTENANCE REPORT

A — ADP REPORTING SECTION

FOR USE OF ADP STAFF ONLY

[-N &Ø1

TYPE OF REPORT (CIRCLE ONE ONLY): &Ø2 C P I F

REPORT SERIAL NUMBER: &Ø3

FACILITY	LOCATION	REASON FOR MAINTENANCE	TOTAL MANHOURS Hrs. Mins.	TRAVELLING TIME IN MANHOURS Hrs. Mins.	EQUIPMENT	NUMBER OF EQUIPMENTS
&Ø4	&Ø5	&Ø6	&Ø7	&Ø8	&Ø9	&1Ø

DATE TIME MAINTENANCE STARTS (ACTUAL) Day Month Time	DATE TIME OF SERVICE INTERRUPTION (ACTUAL) Day Month Time	SERVICE DOWN TIME (ACTUAL) Hrs. Mins	TOTAL WORKING TIME (ACTUAL) Hrs. Mins	
&11	&12	&13	&14	&?

B — PLANNED MAINTENANCE WORK SHEET

HAS USER BEEN INFORMED?	DATE TIME MAINTENANCE STARTS (PLANNED) Day Month Time	DATE TIME OF SERVICE INTERRUPTION (PLANNED) Day Month Time	SERVICE DOWN TIME (PLANNED) Hrs. Mins	TOTAL WORKING TIME (PLANNED) Hrs. Mins
B1 Y NA	B2	B3	B4	B5

Equipment Designation	Work Planned	Work Completed	Equipment Adjusted	Working Time	Man Hours	DETAILS	CA 1512 Serial Number

DISTRIBUTION	*CHARACTERS*	*REASON FOR MAINTENANCE CODES*		
Tels HQ — Green	Figure One — I	Planned Preventive — A	Modification by Station Staff — D	Engineering Investigation — G
Division — Red	Figure Nought — Ø	Runway Change — B	Independent Inspection — E	Awaiting Flight Check — H
STO — Blue	Letter I — I	Interruption to Replace Unit — C	Engineering Modification — F	Work done by Non-Tels Staff — J
Equipment Records — Black	Letter O — O			

Name and Grade in Block Capitals *Date*

Checked by	S.T.O.	C.T.O.	Cross Reference	

Fig. 2.5

CA 1512 REVISED APRIL 1970	NATIONAL AIR TRAFFIC CONTROL SERVICE	EQUIPMENT REPAIR REPORT

FOR USE OF ADP STAFF ONLY

<N &Ø1 | REPORT SERIAL NUMBER &Ø2

LOCATION OF WORKSHOP	HAS CA 1511 BEEN RAISED?	WORKING TIME IN MANHOURS Hrs. Mins	EQUIPMENT
&Ø3	&Ø4 Y N	&Ø5	&Ø6

SYMPTOMS OF FAULT OR WORK TO BE DONE

Name and Grade in Block Letters .... *Date* ....

EQUIPMENT DEFECT SECTION

DESIGNATION	ASSEMBLY	SUB ASSEMBLY	ITEM	ACTION		
&Ø7	&	&	&	&	&	&
&Ø8	&	&	&	&	&	&
&Ø9	&	&	&	&	&	&
&1Ø	&	&	&	&	&	&
&11	&	&	&	&	&	&
&12	&	&	&	&	&	&
&13	&	&	&	&	&	&
&14	&	&	&	&	&	&
&15	&	&	&	&	&	&
&16	&	&	&	&	&	&
&17	&	&	&	&	&	&
&18	&	&	&	&	&	&
&19	&	&	&	&	&	&

&2Ø &?

ACTION CODES	
Replaced, Open Circuit	A
Replaced, Short Circuit	B
Replaced, Intermittent	C
Replaced, Overheated	D
Replaced, Out of Tolerance	E
Replaced, Other Reason	F
Adjusted	G
Repaired	H
Wiring Fault Repaired	J
Cleaned	L
Item Dried Out	K
Lubricated	M
Rubber Moulding Perished	N
No Fault Found	W
Refurbished	X
Sent to Industry	Y
Scrapped	Z

COMMENTS OR WORK CARRIED OUT

CHARACTERS	
Letter O	O
Letter I	I
Figure Nought	Ø
Figure One	1

LOCATION OF EQUIPMENT	HOURS RUN	ITEM SERIAL NUMBER
A	B	C

Name and Grade in Block Letters .... *Date* ....

DISTRIBUTION:- Tels HQ - - Green; Division - - Red; STO - - Blue; Equipment Records - - Black

CROSS REFERENCE

Fig. 2.6

3

Design Factors Determining Down Time

The two main factors governing down time are the mechanical and electrical design of the equipment and the maintenance philosophy. In general, it is the active elements of repair which are determined by the design and the passive elements which are governed by the maintenance philosophy. As has been pointed out in the previous chapter, passive times may be lengthened by an increase in active time because of the need for rest and meal breaks and the associated increase in administrative and logistic delays. The design parameters affecting repair times are discussed below.

Access

Low-reliability parts should be the most easily accessible both in terms of fasteners and covers and in terms of the position of mounting relative to other parts. For example, lamps and very high failure rate devices should be easily removable with the minimum of disturbance to the equipment, and there should be adequate room for the technician to withdraw the device without striking other parts or the chassis. On the other hand, the technician should be discouraged from removing and checking easily exchanged items as a substitute for the correct diagnostic procedure. The socket should be easily visible so that the replacement item can be simply introduced. All wiring and printed boards should be accessible with the minimum of removable covers.

Captive screws and fasteners are highly desirable since they are faster to use and eliminate the risk of losing screws in the equipment—potentially fault producing. Standardization of fasteners and covers leads to familiarization on the part of the technician and hence speedier access. The use of outriggers, which enables printed circuit boards and similar units to be tested whilst still connected to the system, can reduce diagnosis time. On the other hand, such on-line diagnosis could result in maintenance-induced failures and is not always encouraged. In general it is a good thing to minimize the amount of on-line interference with the system by employing easily interchanged units together with alarms and displays providing diagnostic information and easy identification of the faulty unit.

Every least replaceable assembly (LRA) should be capable of removal without the removal of any other LRA or part. The size of the LRA has some effect on the speed of access. The aim is for speedy access consistent with minimum risk of incidental damage.

Adjustment

The amount of adjustment required both during normal system operation and after the replacement of an LRA can be minimized by generous tolerancing during design which will aim at low sensitivity to the drift of characteristics.

Where adjustment is by screwdriver or other tool, care should be taken to ensure that damage cannot be done to the equipment. Guide holes, for example, can prevent a screwdriver from slipping.

Where adjustment requires that some measurements be made, or some indicators be observed, during the adjustment process then the displays or meters should be visible while the adjustment is being made.

It is usually essential for adjustments and alignments to be carried out in a set sequence, and this must be specified in the maintenance instructions.

Built-in Test Equipment

As with all test equipment, built-in test equipment (often abbreviated to BITE) ought, ideally, to be an order of magnitude more reliable than the system of which it is part in order to minimize the incidence of false alarms and false diagnosis or incorrect alignment or even failure to register a fault condition. All these events will either prolong down time or reduce the reliability of the system, and therefore built-in test equipment of poor reliability can grossly reduce the system effectiveness.

The number of physical connections between the system and the built-in test equipment should be minimized to reduce the probability of system faults induced by faults on these connections.

Ideally, built-in test equipment should also be more maintainable than the parent system.

Built-in test equipment carries the disadvantages of being costly, inflexible (since it is designed around the system, it is very difficult to modify) and of requiring some means of self-checking. Furthermore, it adds a weight, volume and power supply penalty to the system. On the other hand, it greatly reduces the amount of realization, diagnosis and checkout time involved in a repair or preventive maintenance activity.

Circuitry

It is not too early to consider maintainability when designing and laying out circuitry. In most cases it is possible to identify a logical sequence of events or signal flow through a circuit, and fault diagnosis is considerably facilitated by a layout of components which reflects this logic. Mechanical considerations of component layout are also important as, for example, in the mounting of fragile components as far as possible from handles or grips. Components should not be so close together as to make damage likely when removing and replacing a faulty item.

The use of microelectronics and integrated circuits introduces difficulties. Their small size together with a large number of leads makes it necessary for connections to be small and close together. This increases the likelihood of damage to components and interconnections during maintenance. The integrated circuit is a functional unit in itself and circuit layout is less able to reflect circuit logic. Since the minimum size of the LRA is decided by equipment practices and methods of interconnection then the LRA will not decrease in size indefinitely. The introduction of integrated circuits has the effect of making the LRA more expensive and include more of the circuitry. Field maintenance at circuit level is more difficult because of increased complexity within a given volume.

Connections

Connections present a classic trade-off between reliability and maintainability. The following types of connection are ranked in order of reliability starting with the most reliable. An approximate idea of relative failure rate is indicated by an index shown against each type. The most reliable is shown as unity.

Wrapped joint . . .	1
Welded connection . .	3
Machine-soldered joint . .	7
Crimped joint . . .	8
Hand-soldered joint . .	10
Edge connector (per pin). .	30

A further factor in the reliability of connections is that, in general, the smaller the connection the less reliable it will be.

Since edge connectors are less reliable than soldered joints, there is a balance between the situation of a few expensive and large plug-in units and that of many smaller throw-away units with the attendant reliability problem introduced by the additional edge connectors. Boards with wrapped joints rather than edge connectors are an order of magnitude more reliable from the point of view of the connections, but the maintainability penalty can easily outweigh the reliability advantage. Consider the time taken to make ten or twenty wrapped joints compared with the time taken to plug in a board equipped with edge connectors.

The following times for making the various types of joint apply to the situation where the appropriate tools (e.g. wrapping tool, soldering iron) are available and prepared.

	seconds
Edge connector (multi-contact)	12
Soldered joint (single-wire)	20
Wrapped joint	54

As can be seen, in terms of maintainability, the various joints are ranked in the opposite order to their reliability.

In general, within the LRA, a high-reliability connection is required and maintainability is a secondary consideration. Here the wrapped or soldered joint is appropriate. The interface between the LRA and the system, on the other hand, requires a high degree of maintainability, so that the plug-in or edge connector is justified. Alternatively, a very high reliability LRA, which is unlikely to require replacement, could justify connection to the system by wrapped joints, whereas a low reliability LRA could be connected to the system by the less reliable plug and socket for quick exchange.

The reliability of a soldered joint (both hand and machine made) is sensitive to the effectiveness of control in the manufacturing process.

Where cable connectors are used it should be ensured by means of labelling, that plugs will not be wrongly inserted in their sockets or inserted in the wrong socket. Mechanical design should be such as to minimize the possibility of damage to connectors or pins by clumsy insertion of plugs or cards.

Where several connections are to be made from one unit or rack to another, the complex of wiring is often supplied to the site as a cableform, the numerous connections being made, at each end of the form, according to some wiring document. It is good maintainability to regard the cableform as an LRA and not to attempt local repairs to a multiway cable. Where the cables are permanently terminated, a faulty wire may be cut back, but left in place, and a new single wire added to replace the link.

Displays and Indicators

Displays and indicators may be effective in reducing diagnostic, checkout and alignment elements of active repair time. Simplicity should be the keynote, and a "go, no go" type of meter or display should require no more than a glance. The use of stark colour changes or other obvious means to divide a scale into areas of "satisfactory operation" and "alarm" condition should be used. Sometimes a meter, together with a multiway switch, is used to monitor several units or conditions in different parts of the same system. Under these circumstances it is desirable that the required indication be the same for all applications of the meter so that a satisfactory condition is shown by little or no change in the indication as the instrument is switched to the various parts of the system. Displays should never be positioned where it is difficult, dangerous or uncomfortable to read them, but this fairly obvious point is often overlooked.

For an alarm condition, audible as well as visual displays are desirable to draw the attention of the technician to the fault. Displays in general, but those relating to an alarm condition in particular, must be more reliable than the system itself, since a failure to indicate an alarm condition is highly undesirable. An unfortunate aspect of alarms is that their failure is likely to conceal dangerous conditions.

If an equipment is unattended then certain alarms and displays may have to be extended to some other location, and the reliability of the transmission medium becomes important to the maintainability and hence availability of the system.

The following points concerning the display of quantitative information are worth noting:

1. False readings can result from parallax effects due to the scale and pointer of a meter being in different planes. A mirror behind the pointer, in some instruments, helps to overcome this difficulty.
2. Where a range exists outside which some parameter is unacceptable then either the acceptable or the unacceptable range should be coloured or otherwise made readily distinguishable from the rest of the scale (Fig. 3.1(*a*)).
3. Where a meter displays a parameter which should normally have a single value then a centre-zero instrument can be used to advantage, and the circuitry arranged for the normal acceptable range of values to fall within the mid-zone of the scale (Fig. 3.1(*b*)).
4. Linear scales are easier to read and less ambiguous than logarithmic scales, but consistency in the choice of scales and ranges minimizes the possibility of misreading (Fig. 3.1(*c*)). On the other hand, there may be occasions when the use of a nonlinear response or false-zero type of meter is positively desirable.
5. Digital displays are now widely available and are, from a maintainability point of view, superior to the analogue pointer-type of instrument where a reading has to be recorded (Fig. 3.1(*d*)). The analogue type of display is preferable when a check or adjustment within a range is required.

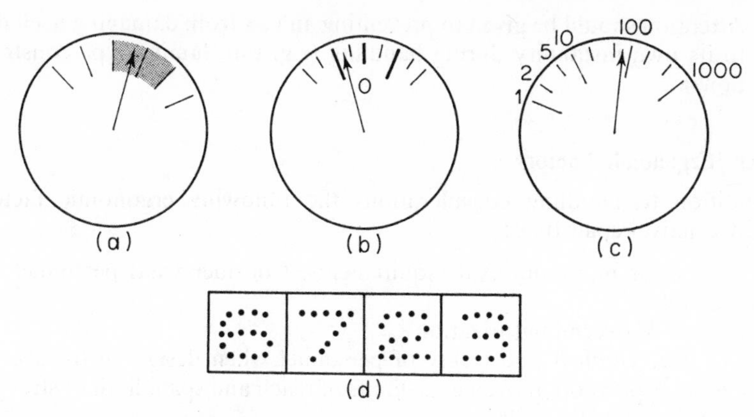

Fig. 3.1 Meter displays

(*a*) Scale with shaded range
(*b*) Scale with limits
(*c*) Logarithmic scale
(*d*) Digital display

Consistency in the use of colour codes, symbols and labels associated with displays should not be overlooked. Filament lamps are not particularly reliable, and where display reliability is important derating and/or duplication may be necessary.

All displays should be positioned as near as possible to the location of the function or parameter to which they refer and might well be mounted in an order comparable to the sequence of adjustment. Unrequired displays merely complicate the situation and do more harm than good.

Handling

Major points to watch are:

1. Weight of removable modules.
2. Size and shape of removable modules.
3. Protection of sharp edges and high-voltage sources. Even an unplugged module may hold dangerous charges on capacitors.
4. Correct handles and grips reduce the temptation to use components for the purpose of holding a module.
5. When an inductive circuit is broken by the removal of a unit then the earth return should not be via the frame. A separate earth return via a pin or connection from the unit should be used.

6. Attention should be given to preventing an LRA from damaging itself due to its own instability during handling (e.g. thin lamina-type constructions).

Human/Ergonomic Factors

In addition to handling considerations the following ergonomic factors influence active repair time:

1. Design for minimum skill requirements. Consider what personnel are available.
2. Beware of overminiaturization.
3. Consider comfort and safety of personnel when designing for access; e.g. body position, movements, limits of reach and span, limit of strength in various positions, etc.
4. Illumination: fixed and portable.
5. Shield from environment (weather, damp, etc.) and from stress generated by the equipment (heat, vibration, noise, gas, moving parts, etc.).

The ergonomic implications of all the other design parameters in this chapter should not be overlooked.

Identification

Identification of components, terminals, leads, connectors and modules is helped by standardization of appearance. Colour codes should not be complex since over 5% of the male population suffer from some form of colour blindness. Simple, unambiguous numbers and symbols help in the identification of particular functional modules. The physical grouping together of functions simplifies the signs required to permit the identification of a particular circuit or LRA.

Interchangeability

Where LRAs are interchangeable this greatly simplifies diagnosis, replacement and checkout due to the element of standardization introduced into the repair action. In addition, spares provisioning becomes slightly less critical in view of the possibility of using a non-essential, redundant, unit to effect a repair in some other part of the system. Cannibalization of several failed LRAs to yield a working module also becomes possible. A greater degree of standardization of parts is introduced into the system as a whole, thereby improving reliability and reducing the cost of component procurement.

The smaller and less complex the LRA the greater are the possibilities of standardization and hence of interchangeability. The penalty lies in the number of interconnections, in total, between LRAs and the system (decreases reliability) and the fact that diagnosis is referred to a lower level (more skill and test equipment).

Interchange of non-identical boards or units should be made mechanically impossible. At the least, circuit/pin conventions should be arranged so that incorrect insertion of a board does not lead to any damage either to the board or to the parent equipment (e.g. similar power supplies always on the same pin number).

Least Replaceable Assembly

The LRA is that replaceable module at which local fault diagnosis ceases and direct replacement from spares takes place. A failure is traced only to the LRA, which should be easily removable from the system (e.g. printed board), replacement LRAs being carried as the unit spare part. It should rarely be necessary to remove an LRA in order to prove that it is faulty, and no LRA should require the removal of any other LRA in order to carry out a replacement.

The choice of level of LRA is one of the most powerful factors in determining the maintainability of an equipment. The larger the LRA unit the faster are the diagnosis and repair of a system failure. Maintainability, however, is not the only factor in the choice of LRA. As the size of LRA increases so does its cost and hence the cost of carrying spare parts. The more expensive an LRA the less likely is a throwaway policy to be applicable. Furthermore, the larger the LRA the less easy it is to achieve interchangeability. The following table illustrates the major considerations in determining the optimum size of a replaceable module.

Size of LRA *increases*	
Maintainability	improves
Reliability (frequency of repair determined by this)	decreases
Cost of testing system (in terms of equipment and manpower	decreases
Cost of spares	increases
Cost of LRA	increases

The optimum combination of the above factors coincides with some particular size of module for a given maintenance environment, labour cost, manpower ability, reliability/maintainability priority, etc. A comparison between alternatives requires some trade-off technique enabling the designer to predict the outcome of the various combinations considered. This will be discussed in a later chapter.

Lubrication

External access to lubrication points speeds up preventive maintenance and reduces the possibility of damage.

Mounting

If components are mounted so as to be self-locating then replacement is made easier. Mechanical design and layout of mounting pins and brackets can be made to prevent transposition where this is undesirable as in the case of a transformer which must not be connected the wrong way round.

Part Selection

Main factors to consider here are:

Availability of spares—delivery
Reliability/deterioration under storage conditions
Ease of recognition
Ease of handling
Cost of parts
Physical strength and ease of adjustment

Redundancy

Circuit redundancy within the LRA (usually unmonitored) increases the reliability of the module, and this technique may be used in order to make the LRA sufficiently reliable to be regarded as a throwaway unit.

Redundancy at the LRA level permits redundant units to be removed for preventive maintenance or for on-the-job training without removing the system from service and with little additional risk of system failure resulting from this temporary loss of a redundant function.

Although improving reliability, and in most cases maintainability, redundancy carries the penalty of added space, weight and cost, and furthermore the additional units generate a need for more spares and more maintenance activities. Although system availability is improved, the number of preventive and corrective maintenance actions must increase with the number of units.

Safety

Apart from legal and moral considerations, safety hazards increase active repair time by requiring greater care from the technician. Furthermore, an unsafe design will encourage him to take short cuts or to omit essential maintenance activities. Accidents add, very substantially, to the repair time.

Where redundancy is present, routine maintenance activities can be carried out after isolation of the unit from high-voltage and other hazards. In some cases maintenance actions require the unit to be live, in which case, appropriate safeguards must be incorporated in the design.

The following practices should be considered as part of normal good design:

1. Isolate high voltages under the control of microswitches which are automatically operated during access. The use of a positive interlock should bar access unless a safe condition has been achieved.
2. Weights should not have to be lifted or supported particularly if danger results.
3. Use appropriate handles which encourage correct handling.
4. Provide physical shielding of high-voltage points, heat and other hazards.
5. Eliminate sharp points and edges.
6. Install adequate alarm arrangements to signal any dangerous system state. The exposure of a distinguishing colour when safety covers have been removed is good practice.
7. Ensure adequate lighting.

Standardization

From the point of view of maintainability, standardization leads to improved familiarization with components and units and hence faster maintenance. The number of different tools and test equipments is reduced, and so also is the possibility of delay due to having the wrong test gear. Fewer types of spares are required thereby reducing the likelihood of a stockout.

Test Points

Test points provide an interface between test equipment and the system for purposes of diagnosis, adjustment and checkout, calibration, monitoring for degradation of performance and for routine checks on redundant units.

The provision of test points is largely governed by the level of LRA chosen for the system and will usually not extend beyond what is required to establish that a particular LRA is faulty.

In order to minimize the possibility of faults being caused as a result of maintenance activities, test points should be positioned within the circuit and buffered by capacitors, resistors, transistors, etc., wherever possible, so as to protect the system from misuse of test equipment. Standardization of appearance and positioning of test points also reduces the possibility of misuse and of incorrect diagnosis and alignment as a result of mistaken connections.

The reliability of all test equipment and circuitry provided, in addition to system functional requirements, for purposes of maintenance should be an order better than that of the system. Incorrect diagnosis caused by faulty test circuitry cannot be permitted.

Since the location and quantity of test points depend on the type of test equipment, the maintenance skills available, the maintenance instructions and the level of diagnosis required, these factors must all be considered at an early stage in the design.

As with displays, additional unnecessary test points are likely to reduce rather than increase system effectiveness.

The above 18 electrical and mechanical design parameters are aspects of the system or the equipment and not of the maintenance organization or arrangements. Their main influence is on the active repair time elements of diagnosis, replacement, checkout, and so on. Design and maintenance philosophies are, nevertheless, interdependent. It was seen that most of the parameters have a strong influence on the choice and design of test equipment, and that attention to one could be traded off against attention to another. The level of human skill is dictated by the level of LRA, types of displays, standardization, etc., and maintenance procedures are influenced by the size of equipment modules, number of different types of spare to be carried, and so on.

Maintenance philosophy also influences active repair time and, in addition, determines the passive time elements.

Summary

Design factors influence active repair time elements and are as follows:

Access
Adjustment
Built-in test equipment
Circuitry
Connections
Displays and indicators
Handling
Human and ergonomic
Identification
Interchangeability
Least replaceable assembly
Lubrication
Mounting
Part selection
Redundancy
Safety
Standardization
Test points

4

The Effect of Maintenance Philosophy on Down Time

Both active and passive repair times are influenced by maintenance procedures, personnel, provisioning of spares and other factors not attributable to the design of an equipment. Consideration of these factors is known as the *maintenance philosophy*, which plays an important part in determining the system effectiveness. The costs involved in these factors are often considerable, and it is therefore important to strike a balance between over- and underemphasizing them. The factors can be grouped under six headings:

Maintenance procedure
Tools and test equipment
Personnel selection, training and motivation
Maintenance instructions, manuals
Spares provisioning
Logistics

Maintenance Procedure

Effective, error-free maintenance carried out in minimum time is best achieved if a logical and formal procedure is followed on each occasion. A haphazard approach based on the subjective opinion of the maintenance technician, although occasionally resulting in spectacular short cuts, is unlikely to prove the better method in the long run. A formal procedure ensures that essential calibration and checks are not omitted, that diagnosis follows a logical sequence designed to prevent incorrect or incomplete fault detection, that correct test equipment is used for each task (damage is likely to be caused by application of incorrect test gear), and that dangerous practices are avoided. Correct maintenance procedure can only be ensured by means of accurate and complete manuals and thorough and effective training of the technician. A maintenance procedure must consist of the following three main parts:

Making and interpreting test readings
Isolating the cause of a fault
Adjusting to ensure optimum performance
Part replacement

The level of fault identification (LRA) determines the extent of the diagnosis, and a number of procedures are used:

1. Stimuli-response, where the response to changes of one or more parameters is observed and compared with the expected response.
2. Parameter checks, where parameters are observed at displays and test points and compared with expected values.
3. Signal injection, where a given pulse or frequency is applied to a particular point in the system and the signal observed at various points in order to detect where it is incorrectly processed or where it is lost.
4. Functional isolation, wherein signals and parameters are checked at various points in a sequence designed to eliminate the existence of the fault before or after each point. In this manner the location of the fault may be narrowed down by a process of systematic elimination.

Having isolated the fault a number of repair methods present themselves:

Direct replacement from store
Cannibalize from non-essential parts
Rebuild using simple construction techniques
Self-repair, i.e. failed unit adapts

Depending upon the circumstances and location of a system a corrective repair may be carried out immediately that a fault is signalled; or, on the other hand, repairs may only be carried out at regular intervals with redundancy being relied upon to maintain performance between visits. The choice between these two approaches has no effect on the maintainability of a system but plays a part in determining the reliability. Fig. 4.1 illustrates the effect, on reliability, of varying the time between periodic visits to a system containing redundancy. If repair is carried out, on a redundant unit, immediately that a fault is evident, the MTBF is obtained from the failure rate of the redundant units and the MTTR of each unit.

In either of the above two cases an immediate repair is carried out if sufficient failures occur to cause a system failure. In this case the maintainability of the unattended, periodically inspected system is likely to be governed by the delay in bringing maintenance staff to the site rather than by diagnosis and repair activities.

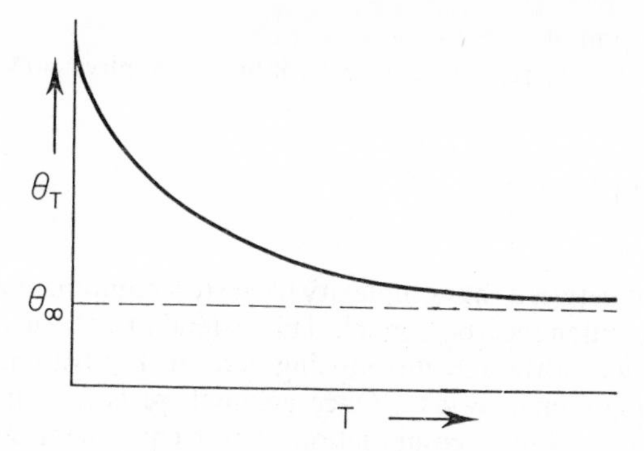

Fig. 4.1 MTBF of periodically inspected system containing redundancy
T = time between periodic visits
θ_T = MTBF
θ_∞ = MTBF when T is infinite

Tools and Test Equipment

The following are a few of the main requirements of test gear:

1. Simplicity: test gear should be easy to use and should not require elaborate set-up procedures and complicated determination of readings.
2. Standardization: the minimum number of types of test gear reduces the training and skill requirement for the technician. It also minimizes the maintenance and spares requirements of the test gear itself. Standardization should extend to the types of display and connections to the equipment.
3. Reliability: test gear ought to be an order of magnitude more reliable than the system for which it is designed since incorrect diagnosis or alignment cannot be tolerated. A test equipment failure can result in a system failure or in unnecessary down time.
4. Maintainability: test gear must be easy to repair and service since its non-availability can lead to system outage.
5. Replaceability: test gear must be capable of being replaced with minimum delay and, if possible, without undue inflation of cost compared with the original equipment.

The same considerations of simplicity, standardization and replaceability apply to tools as well as to test gear. It follows, then, that the main considerations when choosing tools and test equipment are:

Availability and cost of the chosen equipment

The types of existing equipment
Compatibility with the system (environment, test connections, physical size, etc.)
Repair procedures
Number and skill level of personnel
System environment

There is a trade-off situation between the complexity of the test equipment and the skill and training of maintenance personnel. This extends to the use of built-in test equipment which, although introducing certain disadvantages, speeds and simplifies the maintenance task. Once again there is no simple formula for determining the optimum combination of test equipment complexity and manpower ability. The total situation, involving the many variables mentioned, must be considered and a trade-off technique found which takes account of all design parameters together with the maintenance philosophy. Such prediction methods will be discussed in the next chapter.

There is a possibility of automatic test equipment which is separate from the system for which it is designed and which is capable of monitoring several parameters simultaneously or one or more parameters on a repetitive basis by applying appropriate stimuli. On the credit side automatic test equipment can handle more data, and at a faster rate, than simple manual equipment, and can perform more complex tests without imposing special requirements on the maintenance personnel. The possibility of human error is reduced by this type of test gear. These advantages have to be weighed against the fact that automatic test equipment is completely dependent on the system design and is therefore sensitive to any design changes or modifications. Its complexity introduces a reliability problem together with the fact that its maintainability must be of a high order. Automatic test equipment is only capable of carrying out the tests for which it is programmed (built in to its design) and cannot be used for more flexible test decisions. It is, in general, more expensive than standard test equipment and should therefore only be considered for applications where a high utilization can be anticipated. The main factors in determining whether or not to use such test equipment are the degree of utilization, the nature of the tests (repetitive, etc.), and the criticality of the maintenance time.

Another possibility is that of built-in test equipment which forms an integral part of the system and usually requires no setting up procedure in order to initiate a test. Similar considerations apply as with automatic test equipment, but in addition, since it is part of the system, weight, volume and power consumption are more important. A customer may sometimes specify such constraints in the system specification (e.g. power requirements of built-in test equipment not to exceed 2% of mean power consumption). Built-in test gear can be in the form of displays of various parameters rather than a stimuli/response diagnostic device. At the other extreme it may consist of a programmed sequence of stimuli and tests which culminate in a printed-out statement of results. It is therefore of use in detecting degradation of performance by indicating drift of one or more parameters.

With simple portable test equipment there is a choice of general-purpose commercially available equipment as against specially designed equipment. The cost and replaceability of general-purpose equipment are probably more favourable than those of the special type. On the other hand, special-purpose equipment can be made simpler to use, and completely compatible with the test points available, and can hence greatly reduce diagnosis times.

In general, the choice of the various types of test equipment involves trading off complexity, weight, cost and design inflexibility, all of which involve cost, with the advantages of simpler and faster maintenance.

Personnel Considerations

Four manpower considerations influence the maintainability of an equipment:

Training given
Skill level employed
Motivation
Quantity and distribution of personnel

The greater the complexity of the design and hence the variety of maintenance actions the more training is required for the technician. Proficiency in carrying out corrective maintenance actions is a combination of both factual knowledge and diagnostic skill. Knowledge can be acquired by direct teaching methods, but skill can only be gained from experience of the situations concerned either in a real or simulated environment. Maintenance training, then, must include considerable experience of practical fault finding on actual equipment wherever possible. Sufficient theory to enable the technician to understand the reasons for his actions and to make inductive diagnostic decisions is required, but an excess of theoretical teaching is both unnecessary and confusing. A balance has to be achieved between the confusion of too much theory and the motivating interest obtained from broadening the technician's understanding.

A problem often encountered with very high reliability equipment is that certain types of failure occur so infrequently that the maintenance technicians have no experience in their diagnosis and repair. Refresher training by means of simulated faults may be essential to ensure the effectiveness and efficiency of the maintenance force in these circumstances.

Training the maintenance technician in more than one area (e.g. electronic and electromechanical maintenance) provides a more flexible work force and so reduces the likelihood of the appropriate technicians being unavailable to deal with a particular failure. Time wasted in changing from one technician to another in the middle of a maintenance action is eliminated, and the transport and travelling time costs are reduced.

In order to achieve a given performance both specified training and a stated level of ability are assumed. A skill level must be described in objective terms of knowledge, manual dexterity, memory, visual acuity, physical strength, inductive reasoning, and so on. Where a supplier agrees to train the customer's maintenance staff to a given level of proficiency then a preliminary test, of a specified nature, may be necessary to establish their suitability.

Well structured training aimed at providing both flexibility and proficiency contributes to the motivation of the maintenance force, since confidence and the ability to carry out a number of tasks increases the desire to demonstrate both speed and accuracy.

Manpower scheduling requires a knowledge of the failure rates of the equipments concerned. Where different types of failure require different repair times then the manhours of maintenance required for each mode of failure have to be calculated separately from a knowledge of the individual failure rates. Increasing the number of maintenance personnel engaged on rectifying a particular fault may reduce the MTTR. If more than two technicians are engaged on the diagnosis and repair of a fault, however, the increase of manpower is unlikely to achieve a significant improvement in maintainability.

Personnel policies are usually under the control of the customer, and therefore close liaison between design engineers and customer is essential before design features relating to maintenance skills are finalized.

Maintenance Instructions

The most vital repair tool available to the technician is the maintenance manual. It must be accurate and complete, and the information must be easily located or the manual is likely to be discarded by the technician as more of a hindrance than a help.

The main use of the manual is as an aid to diagnosis, for which it should outline a logical sequence of tests necessary to identify, by a process of elimination, the causes of a malfunction. Wherever possible, procedures should be self-checking by ensuring that the omission of a test or of an adjustment or alignment will be detected by a later test or action. Functional block diagrams assist in the process of fault finding by providing a logical illustration of the checking sequence. Illustrations of both correct and incorrect conditions (e.g. of waveforms or displays) speed the process of fault cause recognition.

Other features of the manual should be an outline of alignment and adjustment procedures, emphasis on safety hazards and precautions, details of preventive maintenance actions, and also instructions concerning the use of spares and the requirements for failure reporting.

Every effort should be made to anticipate situations where potential damage could occur as a result of either preventive or corrective maintenance, and these hazards should be clearly indicated in the maintenance instructions.

Such activities as searching for dormant faults and degradation conditions, in order to anticipate and prevent causes of malfunction, are time consuming and therefore expensive. If possible such checks should be made where advantage can be taken of the necessity of access to a particular part of the equipment for other essential reasons.

The maintenance technician will not be motivated to use the manual unless it is clear and readable and does, in his opinion, provide real assistance. To this end the manual should contain no unnecessary theoretical information but should confine itself to the practical requirements of maintenance, using diagrams, pictures and schematics. If representative technicians are consulted during the preparation of the manual a better manual should result, and where the participation is known to the maintenance personnel, they may regard it more favourably.

Chapter 9 is devoted to the principles of handbook design and outlines some methods of presenting maintenance instructions.

Spares Provisioning

The two main variables concerning spares that it is desirable to predict are the likely cost of spares used and the number of spares to be provided.

The number of identical items for which a particular spare is to be kept, together with a knowledge of the failure rate of that item, will yield the likely number and hence cost of that particular type of spare for a given length of time. Since the failure rate is a statistical quantity the cost of spares obtained by this method will not be an exact forecast but another statistical statement. Allowance must be made for spares which may become damaged in transit or due to handling and for those which might fail in store. Also spares may be used to replace parts which are assumed to have failed but are, in fact, fully operational.

The number of spares to be provided, which is related to the risk of a stock-out, requires a knowledge of the statistical distribution of the failures. Assuming a random distribution of failures and hence a constant failure rate, the following method of forecasting spares requirements may be applied. For one particular spare part type, let

λ = failure rate
n = Number of parts of that type which may have to be replaced
r = Number of spares of that type carried

If it is assumed that the failures are randomly distributed then Poisson's formula applies and the probability of r or fewer failures in time t is

$$P_{(0-r)} = \sum_{c=0}^{c=r} \frac{(n\lambda t)^c}{c!} \exp -n\lambda t$$

In other words, this expression gives the probability of no stockout if r spares are carried under the assumption that spares are not replenished during time t.

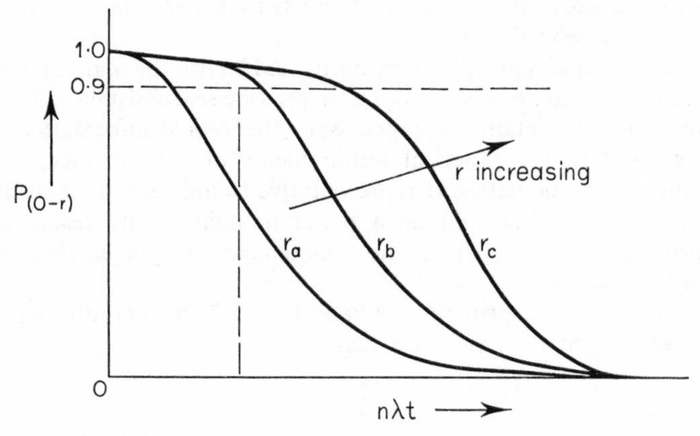

Fig. 4.2 Set of curves for spares provisioning

Fig. 4.2 shows how P_{0-r} varies with $n\lambda t$ for different values of r. Suitably constructed curves of this type or tables of the cumulative Poisson distribution can be used to evaluate the number of spares required for chosen risks of stockout.

Fig. 4.2 shows how a specific value of $n\lambda t$ associated with a 10% risk of stockout yields a spares requirement of r_b.

Another situation occurs where the stock of spares is continually replenished by repaired failed items. Under these circumstances the mean repair time of the failed item is relevant to the calculation. Here repair time refers to the second-line maintenance action of repairing the item and not to the repair time of the system. It is assumed that the system repair is effected by rapid replacement of a failed unit or part and that the failed unit is then repaired and returned to the stock of spares. Fig. 4.3 illustrates this state of affairs.

The following summarizes the main considerations applying to spares provisioning.

1. Failure rate: determines quantity required.
2. Probability of stockout: fixes level of spares.

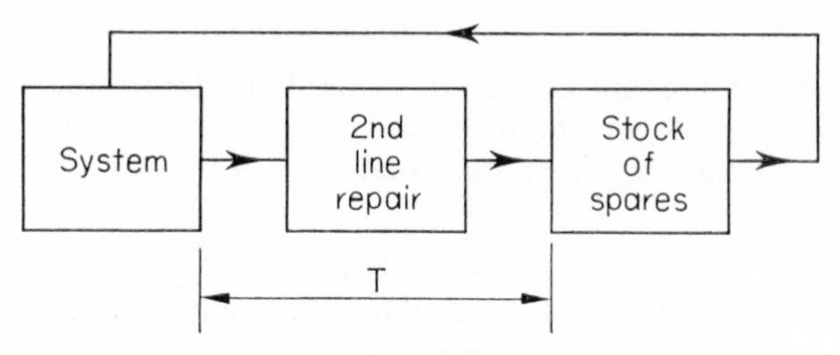

Fig. 4.3 Spares replacement from second-line repair

3. Second-line time to repair: where spares are returned to stock this determines spares level.
4. Cost of spares: the greater the cost of the spare part the greater is the capital tied up in a given number of spare parts. This may set a limit to the number of parts which can be carried.
5. Standardization: the greater the number of different types of spare required the greater is the total capital required to be carried in case of failure.
6. Availability: if spares are ordered as required either for immediate use or to replenish the spares stock then the availability and lead time of ordering are likely to effect the possibility of a stockout and hence to increase the system MTTR.
7. Localization: if spares are carried in a number of small quantities, each batch serving a particular system or part of a system, then the total number of spares overall is greater than that required to form a single central depot. This is simply demonstrated by means of the exponential calculation above. On the other hand, a single depot is likely to give rise to a greater average transport time for the spares when they are required.

The question arises as to whether spares that have been repaired should be returned to some central stock or retain their identity for eventual return to their original system, location or particular equipment position. Return to the exact position occupied previous to malfunction is not possible if an immediate replacement has been carried out unless the replacement is regarded as temporary. It is desirable to retain this separate identity when maintenance is being provided to a number of systems or equipments belonging to different customers or where the alignment of the unit is complex and semipermanent. Another reason for retaining the identity of a particular unit removed for repair is that, where preventive maintenance applies (particularly replacement or attention before wearout), the unit may be replaced by another at a different stage in its life. Both reliability information and preventive maintenance schedules would be adversely effected.

Logistics

Logistics is concerned with the time and effort involved in transporting manpower, equipment and spares to their point of application in order to carry out maintenance actions. The main consideration is that of the degree of centralization of these facilities.

Centralize	*Decentralize*
Special-purpose test equipment	Small tools and standard items
Manpower and test equipment subject to low utilization	Where small MTTR is of overriding importance
Personnel requirements and support	Where transport creates a reliability problem for the test equipment
Second-line repair facilities	

The ideal combination is that where a minimum of these facilities is available at or near the point of application in order to ensure restoration of operational effectiveness in the optimum time, and the remainder of the spares and facilities, including those required for the less probable types of failure, are centralized.

Summary

Maintenance philosophy influences both active and passive repair time elements and covers the following areas:

Maintenance procedure
Tools and test equipment
Personnel: selection, training and motivation
Maintenance instructions: manuals
Spares provisioning
Logistics

5

Predicting Repair Times

The equipment user has long been aware of the high proportion of his costs which are spent on maintenance and provision for maintenance. He is now realizing that the design of equipment plays an important part in determining these costs, and that design and user maintenance practice should be correlated. The designer is forced, either by reason of competition or contract provision, to quantify the maintainability of his equipment to assure himself of a reasonable expectation of meeting the customer's requirements and of doing so at minimum cost.

Active repair times are determined by design, by the maintenance philosophy (Chapters 3 and 4), and by the maintenance instructions, provided that these are carried out by personnel of the required competence and to the letter. Passive repair times depend largely on the user's administrative arrangements and on his policy regarding location of personnel, spares and test gear and his transport arrangements.

All estimating or prediction procedures rely on an element of skill or art as well as a basis of scientific reasoning, and are therefore not 100% justified on the basis of logic. It is all too easy to cloak the subject in an aura of academic respectability and then to be disappointed because practical essentials were overlooked. Those who ignore the practicalities of the actual ironmongery and seek only to produce a few figures acceptable to their management may be tempted to use and believe in a bald mathematical approach.

Any realistic maintainability prediction procedure must meet the following essential requirements:

1. The prediction must be fully documented and described. It must be subject to recorded modification as a result of specific experience. Only in this way can the process of prediction be improved with the knowledge gained from demonstrations and field reports.
2. All assumptions must be recorded and their validity checked where possible.
3. The prediction must be carried out by engineers who are not part of the immediate design group and who therefore are not biased by declared

objectives. The assurances of the design team must be checked, as far as possible, by reference to documentation and physical realizations.

Prediction, valuable as it is, should be replaced by demonstration or by assessment whenever and as soon as possible. Since reliability and maintainability are interrelated, an iterative process will often be necessary. Since the maintainability prediction will always depend on a knowledge of detail failure rates, the results can be no more accurate than the accuracy permitted by the reliability information.

Maintainability prediction is not as well developed as reliability prediction, and it is suggested that the reader should choose one of the published procedures and adapt and evolve it for his own purposes. Two of the procedures from US Military Handbook 472, *Maintainability Prediction*, will be discussed in this chapter and the use of one of them will be illustrated in Chapter 6. Full details of the procedures will not be given here since the object is simply to describe and comment on the approach.

It must be emphasized that these procedures for maintainability prediction are applicable only to the active elements of the maintenance time, since it is these only for which the designer can accept responsibility. This point is not always made clear in the handbook, and Procedure 1 (not described here) includes as active some elements which we have called passive.

Procedure 2: US Military Handbook 472

This procedure was developed by ITT (International Telephone and Telegraph Corporation) for the United States Navy: it is intended for shipboard and shore electronic equipment, and could also be developed for mechanical systems. The procedure is intended for use during the final design stage and is a detailed work study process capable of giving accurate predictions provided that the necessary funds and data are available. It could be adapted and used where progressive developments were being made on a series of equipments and where full data were collected and analysed from the equipments in use. It seems unlikely that it could be applied to equipments which break new ground or where the funding of the maintainability programme is less than lavish. It is likely that, in view of their age, the data given in the handbook would be unsuitable for current designs.

There is little doubt that this method represents a comprehensive, fully accountable and justifiable approach and is as near to an ideal prediction process as is possible. In Europe it will only find application for systems in multiple use by a single operational organization which is sufficiently motivated to fund it over a long term of continuing development. The procedure discusses, in detail, the relationship between *equipment repair time*, ERT, which is defined as the median of the individual repair times, and the *mean time to repair*, MTTR, for different distributions of repair times. Since the procedure is comprehensive, all faults to the *least replaceable assembly*, LRA, level being considered, no assumption of distribution is required for the calculation of MTTR. Furthermore, the ERT could be obtained from the figures listed or from any assumed distribution justified by examination. The forecast of repair time distribution could be a great asset if the equipment considered is to be subjected to a maintainability demonstration test with specified requirements. This is discussed in Chapter 7.

The following information is essential for the use of Procedure 2:

1. The structure and breakdown, in detail, into groups, units, assemblies, sub-assemblies and components or parts of the equipment. This must be to the level of the LRA.
2. The detailed procedure for the diagnosis of a fault.
3. The repair methods to be used.
4. Sufficient information for a reliability prediction based on a full stress analysis. This must include a knowledge of the thermal distribution within the equipment.
5. Mechanical details of the mountings and assemblies within the equipment.

The procedure breaks the active repair time into

Localization: location of the fault to the extent possible without using external test gear
Isolation: location of the fault to the extent possible using external test gear
Disassembly: access to part or LRA to be removed
Interchange
Reassembly
Alignment
Checkout: all checks necessary to confirm that the system performance meets operational requirements

Worksheets A and B (our Figs. 5.1 and 5.2) are used for the calculation of active corrective maintenance times. In column A of Worksheet A, every LRA capable of a fault condition is entered. If repair is to component level then this will include feed-through terminals, wiring, cables and printed circuit boards even where these have no circuit designation for column B.

Where the mode of failure will affect the time to repair, separate times must be allocated for the respective modes and the appropriate partial failure rates entered in column C. Failure mode is most likely to have some effect on the localization time due to the different symptoms produced. Columns D to K are completed in accordance with the data available, and column L gives the

Worksheet A

Contractor______________ Date__________

Contract No._____________ Sheet______of______

A Part Identification	B Circuit Designation	C Failure Rate	Average Time to Perform Corrective Maintenance Tasks							L M_c	M λM_c
			D Localization	E Isolation	F Disassembly	G Interchange	H Reassembly	J Alignment	K Checkout		
	$\Sigma\lambda$									$\Sigma\lambda M_c$	

Fig. 5.1 US Military Handbook 472: Worksheet A

total maintenance time for the condition being considered in that line. Column M contains the product of the failure rate, for that particular failure mode,

Worksheet B

Contractor__________ Date__________

Contract No.__________ Sheet____ of____

1 Worksheet A Sheet Number	2 $\Sigma\lambda$ Column C Total	3 $\Sigma\lambda M_c$ Column M Total	1 Worksheet A Sheet Number	2 $\Sigma\lambda$ Column C Total	3 $\Sigma\lambda M_c$ Column M Total
Subtotals			Subtotals		

Product failure rate, $\Sigma\lambda$ = Total of column 2 subtotals______

Total repair time per 10^6 hours, $\Sigma\lambda M_c$ = Total of column 3 subtotals______

Fig. 5.2 US Military Handbook 472: Worksheet B

and the time to repair. On each sheet the sum of the failure rates in column C ($\Sigma\lambda$) and the sum of the products in column M ($\Sigma\lambda M_c$) are recorded.

Worksheet B is a summary sheet and is used to obtain $\Sigma\lambda$ and $\Sigma(\lambda M_c)$ for the whole equipment. The mean time to repair is predicted by the formula

$$M_c = \frac{\Sigma(\lambda M_c)}{\Sigma\lambda}$$

Worksheet C

Contractor______________ Date______________

Contract No.______________ Sheet______of______

A Description of Preventive Maintenance Task	B f	C M_p	D fM_p
Σf		ΣfM_p	

Fig. 5.3 US Military Handbook 472: Worksheet C

Worksheets C and D (our Figs. 5.3 and 5.4) are used for the calculation of predicted preventive maintenance time. For these a schedule of tasks will have

Worksheet D

Contractor____________ Date____________

Contract No.____________ Sheet_____of_____

1 Worksheet C Sheet Number	2 $\sum f$ Column B Total	3 $\sum fM_p$ Column D Total	1 Worksheet C Sheet Number	2 $\sum f$ Column B Total	3 $\sum fM_p$ Column D Total
Subtotals			Subtotals		

Frequency of preventive maintenance tasks, f = Total of column 2 sub-totals_____
Total preventive maintenance time per million hrs, fM_p = Total of column 3 sub-totals_____

Fig. 5.4 US Military Handbook 472: Worksheet D

been determined and instructions as to when they will be performed laid down. On Worksheet C, column A is used to describe the task, in column B the frequency is recorded, column C contains the time for the maintenance action (M_p), and column D is for the product of columns B and C. Σf and $\Sigma(fM_p)$ are calculated for the sheet and transferred to Worksheet D. The mean preventive maintenance time for the equipment is predicted from the formula

$$M_p = \frac{\Sigma(fM_p)}{\Sigma f}$$

Procedure 3: US Military Handbook 472

Procedure 3 was developed by RCA (Radio Corporation of America) for the US Air Force and was intended for ground systems. It requires a fair knowledge of the detail of the design and of maintenance procedure for the system which is to be analysed. The method is based on the principle of predicting a sample of the maintenance tasks. It is entirely empirical, having been developed to agree with known answers for

1. Long-range search radar with two operational channels and comprising some 3000 components and where repair is carried out to component level.
2. A two-channel data processor with about 115000 components with repair to module level.
3. A time-division data-link transmitter with digital and radio-frequency elements having some 45000 components repaired to module level in the digital area and to component level in the r.f. area.

The handbook records the following correlations:

	Final prediction	*Observed active repair*
AN/FPS—6 Radar	67 min	94 min
AN/GRT/GRR7	52 min	63 min

The correlation for the radar improved when the sample was adjusted to correspond to field failures that had been experienced. The MIL Handbook devotes six pages to the procedure for selecting the sample of failures whose repair times are to be predicted. It is stated that "on the average it should take the same time to correct any resistor or capacitor failure as is required for any other resistor or capacitor." This may not always be true.

Where repair at system level is done by replacement of sizeable modular elements (i.e. a large LRA) and the sample is calculated on the basis of these then the cost in terms of time and effort is low enough to permit the use of an adequate sample. In this case the contribution to the total sample, of each group of similarly identified modes of failure, is determined by the predicted failure rate for each of these failures. This can result in considerable simplification of the process as the example in Chapter 6 will show.

The predicted repair time for each sample task is arrived at by considering a check list of maintainability features and by scoring points according to the degree of conformity with the "optimum" for each feature. The score for each feature increases with the degree of conformity with the ideal. The questions, and hence the points scored, are grouped under three headings, Design, Support and Personnel. The points scored under each heading are appropriately weighted and related to the predicted repair time by means of a regression equation.

Equip.__________ Unit/Part__________ Task No.__________

Ass'y__________ By__________ Date__________

Primary function failed unit/part__________

Mode of failure__________

Malfunction symptoms__________

Maintenance Analysis

Maintenance Steps	Scoring Comments

Checklist Scores

	1	2	3	4	5	6	7	8	9	10	11	12	13	14	15	Total
A																
B																
C																

Predicted downtime ____ Min.

Fig. 5.5 US Military Handbook 472: Maintainability Prediction Form

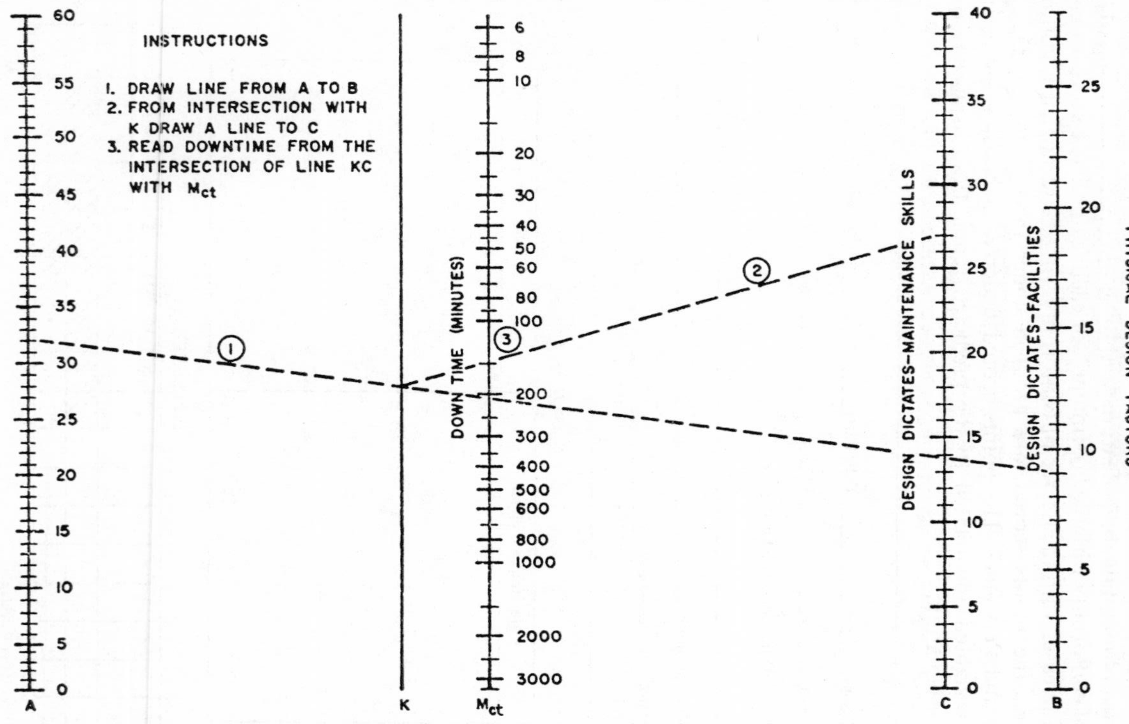

Fig. 5.6 US Military Handbook 472: downtime nomograph

Fig. 5.5 shows the score sheet for use with the checklist, and Fig. 5.6 presents the regression equation in the form of a nomograph. When considering the checklist it will be noticed that additional weight is given to certain features of the design or support by the fact that the answers to more than one question are influenced by a particular single feature.

The checklist is reproduced, in part, in Chapter 6 for the purposes of the example, but the reader wishing to perform a prediction for himself will need a copy of US Military Handbook 472 in order to obtain the complete list. The application of the checklist to typical tasks is, in the authors' opinion, justified as an aid to maintainability design review, even if a prediction of repair times is not specifically required.

Summary

The essential requirements of a maintainability prediction procedure are:

Full documentation
Assumptions fully explained and justified
Carried out by persons outside the immediate design group

The only available methods are described in US Military Handbook 472. Procedures 2 and 3 of that handbook are described in this chapter.

Procedure 2

Predicts active repair times in the elements of localization, isolation, disassembly, interchange, reassembly, alignment, checkout.

Work study approach which allocates repair times to each element of each repair.

Requires detail knowledge of structure and breakdown of the system to LRA level, diagnosis procedure, repair methods, reliability prediction, mechanical mounting details.

Expensive.

Procedure 3

Predicts total active repair time.

Requires knowledge of mechanical design.

Empirical method based on scoring points according to the degree of conformity with various design criteria.

Not so expensive as Procedure 2.

Worthwhile as a design checklist even if prediction is not specifically required.

6

Maintainability Prediction—A Case Study

This study is based on an actual case arising in the normal course of business. An instrument landing system, of which we will consider a part, had been designed and was in production. The design was aimed at a configuration capable of use in the landing of aircraft under automatic control in conditions of low or zero visibility. The part discussed is the *localizer*, which is capable of giving azimuth (horizontal) guidance to the aircraft onto and along the extended centre-line of the runway and guidance along the runway after touch-down. The design had been carried out with this end in view with the obvious need for high confidence in the accuracy and the continuity of the guidance signal. The illustrations are all of the system as fitted for this critical purpose. The proposed requirement, for which the maintainability prediction was performed, would have been for a less onerous standard of security but for which the establishment and demonstration of system repair time would have been demanded. Assuming the reduction in the monitoring and changeover requirements and certain minor mechanical changes, the prediction indicated that the equipment would have met the standard required with adequate margins and that they could have been accepted with great confidence. Although a formal maintainability programme, as envisaged in this text, had not been performed, the design requirements coupled with a thorough reliability programme had resulted in a design which could fulfil the new maintainability requirements.

A full technical description of the equipment would not be appropriate in this book, and we will therefore give as brief an outline as will serve our purpose. The localizer produces a radio-frequency pattern in space which enables a receiver in the aircraft to identify the aircraft position relative to a vertical plane in space which includes the desired flight and "roll out path" of the aircraft. Deviations to the left or right of this plane are indicated visually to the pilot and electrically to the auto-pilot.

The localizer consists of a *remote control unit* situated in the control tower and accessible to air traffic control, a *main cabinet* including transmitters and monitors, an aerial distribution unit, an aerial array, and a number of monitoring receiving aerials (monitor probes). The aerial is situated beyond the "stop end" and facing along the runway with the aerial distribution unit behind it. The building housing the main cabinet is within about 100 m of the aerial and to the rear. The two sets of monitor probes are distributed in front of the aerial, and some may be discerned in the picture of the aerial. In addition, the ILS building houses batteries, chargers, power supply units, r.f. attenuators and, where necessary, air-conditioning equipment.

Plate 1 shows the remote control unit. A normally operating system shows two unbroken horizontal lines of indicator pattern. The photograph shows an unenergized equipment with all indicators showing the unsatisfactory state except for the one which indicates that the standby transmitter is not radiating, i.e. the normal condition is unenergized.

Plates 2, 3 and 4 show the main cabinet. The common control unit can be seen at lower centre and is shown in greater detail in Plate 5. This unit includes a meter and switches which enable many checks to be made on the transmitters. Panels below and to the side show the readings required. To the left and right of the control unit are the main and standby transmitters, and beneath are the associated mechanical modulator and motor drive unit for each transmitter. The upper half of the cabinet houses four similar groups of units each comprising three monitor units (Plate 6) and their associated alarm unit (Plate 7). Each monitor unit checks three characteristics of the transmitted signal, and hence each group confirms the correct condition of the signal for nine characteristics. The meter on the alarm unit can show the condition of each of the nine characteristics. The aerial array is shown in Plate 8.

In the case illustrated a set of monitors checks the signals produced by the "in-use" transmitter, a second checks the signal observed by one set of monitor probes, and a third the signal observed by the second set of monitor probes. A fourth set of monitors observes the signals being fed to the dummy load by the standby transmitter. In the case studied this set would be omitted, but provision would be made for checking the standby with a portable set of monitoring equipment.

On each side of the cabinet is an r.f. distribution unit associated with the adjacent transmitter and modulator. This unit is a stripline waveguide structure which processes the r.f. carrier before and after modulation to produce the four modulated r.f. signals required.

A third waveguide structure, the *coaxial distribution unit*, is mounted at the top and at the back of the cabinet. It feeds the signals from one transmitter out to the aerial and from the other to a set of dummy loads. It incorporates the r.f. changeover relays and also output for the monitor units.

The automatic changeover of transmitters is effected when alarms are given by two of the three monitor sets at one time. This avoids the risk of an unnecessary changeover as a result of a single fault in the monitoring system.

The extent to which the equipment may be relied upon for landing is precisely scheduled in accordance with the state of warnings and alarms and of the maintenance being performed. Maintenance only takes place with the knowledge and consent of air traffic control.

As a preliminary to the prediction a maintainability study was made. This was required to determine the test equipment necessary and the scales of spares holdings. It also examined the design critically and identified the necessary additional test facilities resulting from the reduction in the built-in monitoring together with certain improvements desirable to aid the attainment of specific repair times. It is of interest that these were mainly in the area of mechanical details which improved accessibility and protected disturbed units. This is something which often receives inadequate attention from electronic engineers. In this case the changes were minor and included the closing of openings through which items could be dropped and thus give rise to extensive disassembly for their removal, the discarding of some covers whose function was entirely aesthetic and the substitution of coaxial plugs and sockets for more permanent connections. The latter is an illustration of a trade-off between reliability and maintainability (see Chapter 3).

It was decided to carry out a prediction in accordance with Procedure 3 of US Military Handbook 472 as described in Chapter 5. This was the only method found to be appropriate since the time available was limited and the necessary data which would have been required for alternative methods were not immediately available.

The sample size was determined as 50. A full stress analysis and reliability prediction had been carried out on the equipment and the expected failure rates of the various elements were known. It was possible, therefore, to describe fairly broad modes of failure for the elements, and an actual sample of 20 representative tasks was identified and their relative frequency of occurrence in the total of 50 tasks was determined to the nearest integer. This was possible because system maintenance was generally on the basis of direct replacement of sizeable units and many units occurred two or more times. It was necessary to select one task only from several units whose individual failure rates would not justify separate tasks, and then to regard this as representative of the whole population of such items.

The tasks identified were as follows:

Task number	*Fault*
1–10	Transmitter fault—no r.f. power
11	Crystal oven—control lost
12	Identity tone faulty
13	Modulator—motor speed wrong

Task number	*Fault*
14	Modulator fault—one r.f. disconnection on the modulator
15	Modulator fault—mechanical misalignment
16–23	Monitor fault—false alarm
24–25	Alarm unit fault—false alarm
26	Alarm unit fails to give alarm
27	Common control fault
28	Remote control fault
29	R.F. distribution unit faulty
30	Coaxial distribution unit faulty—fault of changeover relay
31	Coaxial distribution unit faulty—different fault of changeover relay
32–34	Motor drive unit faulty
35	Monitor preamplifier—no output
36	Fault in wiring on the framework—one specified disconnection
37–48	Charger power unit faulty
49–50	Air conditioning unit faulty

The two tasks illustrated are Tasks 1–10 and 16–23. It should be noted that time awaiting air traffic control permission is passive maintenance time and does not form part of the prediction.

Example 1—Tasks 1–10

Fault
No power output from the in-use transmitter.

Symptom
Three monitor sets alarm—auto changeover—service restored.

Maintenance steps
1. Switch suspect transmitter to dummy load.
2. Operate meter switches on common control unit to confirm that the transmitter is faulty.
3. Replace transmitter.
4. Check on dummy load and then, when permitted, confirm that the monitors accept this new transmitter after changeover.

Table 6.1

Checklist A	*Example* 1 *Transmitter fault*	*Example* 2 *Monitor fault*
1. Access (external)	4	4
2. Latches and fasteners (external)	2	2
3. Latches and fasteners (internal)	4	4
4. Access (internal)	4	4
5. Packaging	4	4
6. Units/parts (failed)	4	4
7. Visual displays	4	4
8. Fault and operation indicators	4	2
9. Test points availability	4	3
10. Test points identification	4	2
11. Labelling	4	4
12. Adjustments	4	2
13. Testing in circuit	4	0
14. Protective devices	4	4
15. Safety—personnel	4	4
	58	47

Checklist B		
1. External test equipment	4	2
2. Connectors	4	2
3. Jigs and fixtures	4	4
4. Visual contact	4	4
5. Assistance operations	2	2
6. Assistance technical	4	4
7. Assistance supervisory	4	4
	26	22

Checklist C		
1. Arm-leg-back strength	4	4
2. Endurance and energy	4	4
3. Eye–hand	4	4
4. Visual	4	4
5. Logic	3	3
6. Memory	3	3
7. Planning	4	3
8. Precision	2	2
9. Patience	4	4
10. Initiative	4	4
	36	35

Example 2—Tasks 12–23

Fault

Monitor gives a warning or alarm due to an internal fault.

Symptom

One alarm or warning only—a manual changeover, carried out when permitted, does not remove the condition.

Maintenance steps

1. Identify monitor as the source of alarm.
2. Remove monitor.
3. Check with portable field test monitor that monitor inputs are, in fact, good.
4. Replace monitor with a spare.
5. Check and carry out required and permitted adjustments.

Table 6.1 shows the scoring for the two illustrated tasks, and these will be discussed together as briefly as is meaningful. In some cases the checked item is not relevant and a maximum score is then given. The first three items of Checklist A, the first two items of Checklist B, and the scoring criteria for Checklist C, in US Military Handbook 472, are reproduced below. The chapter concludes with our comments on the checklist scoring.

Check List A—Scoring Physical Design Factors

(1) *Access* (*External*): Determines if the external access is adequate for visual inspection and manipulative actions. Scoring will apply to external packaging as related to maintainability design concepts for ease of maintenance. This item is concerned with the design for external visual and manipulative actions which would precede internal maintenance actions. The following scores and scoring criteria will apply:

Scores

(*a*) Access adequate both for visual and manipulative tasks (electrical and mechanical) 4
(*b*) Access adequate for visual, but not manipulative, tasks . . . 2
(*c*) Access adequate for manipulative, but not visual, tasks . . . 2
(*d*) Access not adequate for visual or manipulative tasks . . . 0

Scoring Criteria
An explanation of the factors pertaining to the above scores is consecutively shown. This procedure is followed throughout for other scores and scoring criteria.

(*a*) To be scored when the external access, while visual and manipulative actions are being performed on the exterior of the subassembly, does not present difficulties because of obstructions (cables, panels, supports, etc.).
(*b*) To be scored when the external access is adequate (no delay) for visual inspection, but not for manipulative actions. External screws, covers, panels, etc., can be located visually; however, external packaging or obstructions hinders manipulative actions (removal, tightening, replacement, etc.).
(*c*) To be scored when the external access is adequate (no delay) for manipulative actions, but not for visual inspections. This applies to the removal of external covers, panels, screws, cables, etc., which present no difficulties; however, their location does not easily permit visual inspection.
(*d*) To be scored when the external access is inadequate for both visual and manipulative tasks. External covers, panels, screws, cables, etc., cannot be easily removed nor visually inspected because of external packaging or location.

(2) *Latches and Fasteners* (*External*): Determines if the screws, clips, latches, or fasteners outside the assembly require special tools, or if significant time was consumed in the removal of such items. Scoring will relate external equipment packaging and hardware to maintainability design concepts. Time consumed with preliminary external disassembly will be proportional to the type of hardware and tools needed to release them and will be evaluated accordingly.

Scores
(*a*) External latches and/or fasteners are captive, need no special tools, and require only a fraction of a turn for release 4
(*b*) External latches and/or fasteners meet two of the above three criteria 2
(*c*) External latches and/or fasteners meet one or none of the above three criteria 0

Scoring Criteria
(*a*) To be scored when external screws, latches, and fasteners are:

(1) Captive
(2) Do not require special tools
(3) Can be released with a fraction of a turn

Releasing a "DZUS" fastener which requires a 90-degree turn using a standard screw driver is an example of all three conditions.

(*b*) To be scored when external screws, latches, and fasteners meet two of the three conditions stated in (*a*) above. An action requiring an Allen wrench and several full turns for release shall be considered as meeting only one of the above requirements.

(*c*) To be scored when external screws, latches, and fasteners meet only one or none of the three conditions stated in (*a*) above.

(3) *Latches and Fasteners* (*Internal*): Determines if the internal screws, clips, fasteners or latches within the unit require special tools, or if significant time was consumed in the removal of such items. Scoring will relate internal equipment hardware to maintainability design concepts. The types of latches and fasteners in the equipment and standardization of these throughout the equipment shall tend to affect the task by reducing or increasing required time to remove and replace them. Consider "internal" latches and fasteners to be within the interior of the assembly.

Scores

(*a*) Internal latches and/or fasteners are captive, need no special tools, and require only a fraction of a turn for release 4
(*b*) Internal latches and/or fasteners meet two of the above three criteria . 2
(*c*) Internal latches and/or fasteners meet one or none of the above three criteria 0

Scoring Criteria

(*a*) To be scored when internal screws, latches and fasteners are:

(1) Captive
(2) Do not require special tools
(3) Can be released with a fraction of a turn

Releasing a "DZUS" fastener which requires a 90-degree turn using a standard screw driver would be an example of all three conditions.

(*b*) To be scored when internal screws, latches, and fasteners meet two of the three conditions stated in (*a*) above. A screw which is captive can be removed with a standard or Phillips screw driver, but requires several full turns for release.

(*c*) To be scored when internal screws, latches, and fasteners meet one of three conditions stated in (*a*) above. An action requiring an Allen wrench and several full turns for release shall be considered as meeting only one of the above requirements.

Check List B—Scoring Design Dictates—Facilities

The intent of this questionnaire is to determine the need for external facilities. Facilities, as used here, include material such as test equipment, connectors, etc., and technical assistance from other maintenance personnel, supervisor, etc.

(1) *External Test Equipment*: Determines if external test equipment is required to complete the maintenance action. The type of repair considered maintainably ideal would be one which did not require the use of external test equipment. It follows, then, that a maintenance task requiring test equipment would involve more task time for set-up and adjustment and should receive a lower maintenance evaluation score.

Scores

(*a*) Task accomplishment does not require the use of external test equipment 4
(*b*) One piece of test equipment is needed 2
(*c*) Several pieces (2 or 3) of test equipment are needed . 1
(*d*) Four or more items are required 0

Scoring Criteria

(*a*) To be scored when the maintenance action does not require the use of external test equipment. Applicable when the cause of malfunction is easily detected by inspection or built-in test equipment.
(*b*) To be scored when one piece of test equipment was required to complete the maintenance action. Sufficient information was available through the use of one piece of external test equipment for adequate repair of the malfunction.
(*c*) To be scored when 2 or 3 pieces of external test equipment are required to complete the maintenance action. This type of malfunction would be complex enough to require testing in a number of areas with different test equipments.
(*d*) To be scored when four or more pieces of test equipment are required to complete the maintenance action. Involves an extensive testing requirement to locate the malfunction. This would indicate that a least maintainable condition exists.

(2) *Connectors*: Determines if supplementary test equipment requires special fittings, special tools, or adaptors to adequately perform tests on the electronic system or sub-system. During troubleshooting of electronic systems, the minimum need for test equipment adaptors or connectors indicates that a better maintainable condition exists.

Scores

(*a*) Connectors to test equipment require no special tools, fittings, or adaptors 4
(*b*) Connectors to test equipment require some special tools, fittings, or adaptors (less than two) . . . 2
(*c*) Connectors to test equipment require special tools, fittings, and adaptors (more than two) 0

Scoring Criteria

(*a*) To be scored when special fittings or adaptors and special tools are not required for testing. This would apply to tests requiring regular test leads (probes or alligator clips) which can be plugged into or otherwise secured to the test equipment binding post.
(*b*) Applies when one special fitting, adaptor or tool is required for testing. An example would be if testing had to be accomplished using a 10 db attenuator pad in series with the test set.
(*c*) To be scored when more than one special fitting, adaptor, or tool is required for testing. An example would be when testing requires the use of an adaptor and an RF attenuator.

(3) *Jigs or Fixtures*: Determines if supplementary materials such as block and tackle, braces, dollies, ladder, etc., are required to complete the maintenance action. The use of such items during maintenance would indicate the expenditure of a major maintenance time and pinpoint specific deficiencies in the design for maintainability.

Scores

(*a*) No supplementary materials are needed to perform task . 4
(*b*) No more than one piece of supplementary material is needed to perform task 2
(*c*) Two or more pieces of supplementary material are needed 0

Scoring Criteria

(*a*) To be scored when no supplementary materials (block and tackle, braces, dollies, ladder, etc.) are required to complete maintenance. Applies when the maintenance action consists of normal testings and the removal or replacement of parts or components can be accomplished by hand, using standard tools.
(*b*) To be scored when one supplementary material is required to complete maintenance. Applies when testing or when the removal and replacement of parts requires a step ladder for access or a dolly for transportation.

(*c*) To be scored when more than one supplementary material is required to complete maintenance. Concerns the maintenance action requiring a step ladder and dolly adequately to test and remove the replaced parts.

Check List C—Scoring Design Dictates—Maintenance Skills

This check list evaluates the personnel requirements relating to physical, mental, and attitude characteristics, as imposed by the maintenance task.

Evaluation procedure for this check list can best be explained by way of several examples. Consider the first question which deals with arm, leg and back strength. Should a particular task require the removal of an equipment drawer weighing 100 pounds, this would impose a severe requirement on this characteristic. Hence, in this case the question would be given a low score (0 to 1). Assume another task which, due to small size and delicate construction, required extremely careful handling. Here question 1 would be given a high score (4), but the question dealing with eye-hand coordination and dexterity would be given a low score. Other questions in the check list relate to various personnel characteristics important to maintenance task accomplishment. In completing the check list, the task requirements for each of these characteristics should be viewed with respect to average technician capabilities.

Scores

	Score
1. Arm, Leg, and Back Strength	——
2. Endurance and Energy	——
3. Eye–Hand Coordination, Manual Dexterity, and Neatness	——
4. Visual Acuity	——
5. Logical Analysis	——
6. Memory—Things and Ideas	——
7. Planfulness and Resourcefulness	——
8. Alertness, Cautiousness, and Accuracy	——
9. Concentration, Persistence and Patience	——
10. Initiative and Incisiveness	——

Scoring Criteria

Quantitative evaluations of these items range from 0 to 4 and are defined in the following manner:

4. The maintenance action requires a minimum effort on the part of the technician.
3. The maintenance action requires a *below average* effort on the part of the technician.

2. The maintenance action requires an *average* effort on the part of the technician.
1. The maintenance action requires an *above average* effort on his part.
0. The maintenance action requires a *maximum* effort on his part.*

Comments on the Checklist Scoring

Checklist A

Item 1 Access (external) scores a maximum with both examples.

Item 2 Latches and fasteners (external). Captive fasteners with knurled knobs are used which require at least one full turn for release. This gives a score of 2.

Item 3 Latches and fasteners (internal) give a score of 4 in both examples since the simple plug-in units involve no such fasteners.

Item 4 Access (internal) is irrelevant and therefore scores 4.

Item 5 Packaging refers to the disassembly process and is also irrelevant.

Item 6 Units/parts failed refers to the manner in which units or parts are removed. This is not relevant and scores 4.

Item 7 Visual displays. Both examples involve built-in visual displays in the area of the main cabinet and so score 4. The fact that some information must be sought when carrying out Example 2 is covered by other items in the checklist.

Item 8 Fault and operation indicators (built-in test equipment). Example 1 scores 4 since the fault information is presented clearly. In Example 2, although the information is presented clearly, operator interpretation is required and the score is 2.

Item 9 Test points availability. Example 1, requiring no test points, scores 4. Example 2 has test points available for all needed tasks and scores 3.

Item 10 Test points identification. Example 1, requiring no test points, scores 4. Example 2, while having identified test points, does not have the required readings displayed and these could not have been practically provided. It therefore scores 2.

Item 11 Labelling. Clear unit identification is a feature of both examples and 4 is scored.

Item 12 Adjustments. Example 1 requires no adjustment—score 4. Example 2 requires minor adjustments to the replaced unit—score 2.

Item 13 Testing (in circuit). In Example 1 the fault is identified while the unit is in circuit—score 4. In Example 2 the unit requires removal for identification of the fault—score 0.

* This concludes the extract from US Military Handbook 472.

Item 14 Protective devices. In both cases the design ensures self-protection against damage after malfunction, and both examples score 4.

Item 15 Safety—personnel. The examples involve no risk to personnel and therefore score 4.

Checklist B

Item 1 External test equipment. Example 1 requires none and scores 4, whilst Example 2 requires one item and scores 2.

Item 2 Connectors. Example 1 requires none and scores 4, whilst Example 2 requires one and scores 2.

Item 3 Jigs and fixtures are not required and the score is 4.

Item 4 Visual contact. Not relevant where one technician is required, so 4 is scored.

Item 5 Assistance—operations. In all cases clearance to work is required and 2 is scored.

Item 6 Assistance—technical. Only one technician is required, so 4 is scored.

Item 7 Assistance—supervisory. No help is required, so 4 is scored.

Checklist C

This list scores according to the degree to which various human skills and attributes are required for the repair task. It is perhaps the most difficult list to score, and opinions will not always be identical.

Items 1–4 Strength, endurance, eye–hand co-ordination and visual acuity are hardly required for these tasks, and 4 is scored.

Items 5 and 6 A small degree of logical reasoning and memory is required, and both tasks score 3.

Item 7 Example 1 requires no planning and scores 4, whereas Example 2 requires a small amount of planning of the use of test equipment and the carrying out of checks. It therefore scores 3.

Item 8 A certain degree of precision is required for both tasks, and a score of 2 is reasonable.

Items 9 and 10 Neither patience nor initiative are required, and 4 is scored.

Using the score totals from Table 6.1, the reader may refer to the nomograph in Chapter 5 (Fig. 5.6). He will find that the active repair times for the two tasks are: Example 1, 8 min; Example 2, 20 min.

The actual times were not established in practice, but the predictions were considered, by those with experience of this type of equipment, to be realistic. The longest predicted repair time in the sample was 120 min, but this refers to a fault which would not prevent the signal from being maintained in space.

Its effect would only be to reduce the security of the system for the duration of the repair. The longest predicted repair time, in the sample, of a fault resulting in total loss of signal not recoverable by manual control was 26 min.

The authors are aware of this prediction process giving realistic repair time assessments of many hours (not referring to this sample) and consider that it is valid over a range extending from the minimum value of 5 min (all scores of 4) to at least 10 h.

7

Demonstrating Maintainability

Chapters 5 and 6 have discussed the prediction of maintainability during the design stage. Once a prototype or production model has been built, however, the actual maintainability may have to be assessed. This may be for the purpose of providing information and guidance as an input to future designs or for improvements to the system in question; or, on the other hand, it may be for the purpose of establishing compliance with some contractual requirements. In either case the demonstration may be carried out under artificial "test" conditions or it may be based on a record of "in use" conditions.

Where demonstration of maintainability is a contractual requirement it is essential that the method and conditions of the demonstration are included in the contractual document in full detail in order to minimize the likelihood of disagreements arising during the test.

Both the supplier and the customer wish to achieve the maintainability objective at minimum cost, and yet a precise measurement of the achieved level by means of a contractually based test, having acceptable statistical risks of error for both parties, is extremely expensive. Both parties must recognize that this may divert money from the achievement of maintainability and should therefore settle for a compromise test containing realistic probabilities of error.

The only reference document concerning these tests is US Military Standard 471, *Maintainability Demonstration*, 15 February 1966, together with Notice 1 of 9 April 1968. If this document is used as the basis for a contractual agreement then the supplier of the equipment should assess for himself the risks that he will run in any agreed demonstration. If specialist statistical advice is sought to this end, it is important for the engineer to explain fully any assumptions to the statistician and also to ask him the right questions. Statistical mathematics are usually dependent on assumptions which have no established physical justification (e.g. the distribution of repair times), and therefore such assumptions must be recognized and be shown to be sufficiently near to the truth to constitute an acceptable approximation.

A true assessment of the maintainability achieved can only be obtained at the end of the fully observed life of the equipment concerned. Anything less than this is a sample which will not even be random, being constrained in time, place, maintenance personnel and the faults (or simulated faults) observed. In addition to the statistical risks of obtaining answers which are in error beyond the acceptable limits (from a sample of maintenance tasks) there exists the risk of failing to achieve the required objective in practice due to inadequate or incorrect maintenance procedure, physical non-realization of the design, unreliable test equipment, and so on.

The theory which deals with the risks of sampling error borne by the supplier and the customer is the same as that which deals with quality and reliability testing.* This will be briefly outlined here.

Consider a population of items having some parameter or measurement (e.g. repair time) which varies between one item and another. Assume that the population is acceptable to the customer provided that the mean value of this parameter is better than a certain value, X. The customer may agree to accept the population provided that the mean value of some random sample, taken from the population, is better than some value X_s which is usually more favourable than X. The so-called "*operating characteristic*" (OC) of the sampling plan is a curve obtained by plotting the probability of accepting the population against the true value of the mean of the given characteristic for the population submitted. In other words the OC curve of the sampling plan is a curve showing the probability of accepting the submitted lot against different values of the mean of the parameter being inspected.

Should the whole population be precisely measured then the OC curve will be the step function shown by the broken line in Fig. 7.1. In this case the lot is correctly accepted or rejected on the basis of the inspection since there is no sampling risk involved, the inspection being 100%.

In point of fact, since the test is performed on a sample, the probability of passing or failing the test is neither zero nor unity for any finite value of the variable but varies as is shown by the continuous line in Fig. 7.1.

The actual shape of the OC curve is determined by the size of the sample, the statistical distribution of the values of the parameter in the population submitted, the acceptance criteria set, the size of the population, and the randomness of the sample. For any lot submitted the test can only result in acceptance or rejection, and therefore the probabilities of these two possibilities must sum to unity. In order to talk in terms of risks which are acceptable to one party or the other, they must be expressed in terms of different values of the required parameter as follows:

Producer's risk, α, is the probability of rejecting lots with mean value of the parameter or dimension to be checked X_0.

Consumer's risk, β, is the probability of accepting lots with mean value of the parameter or dimension to be checked X_1.

* Smith, D. J., *Reliability Engineering* (Pitman, 1972).

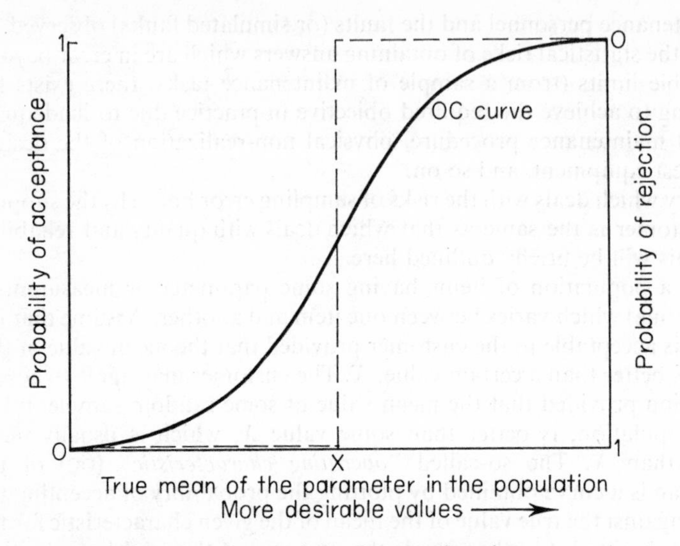

Fig. 7.1 Operating characteristic of a sampling plan

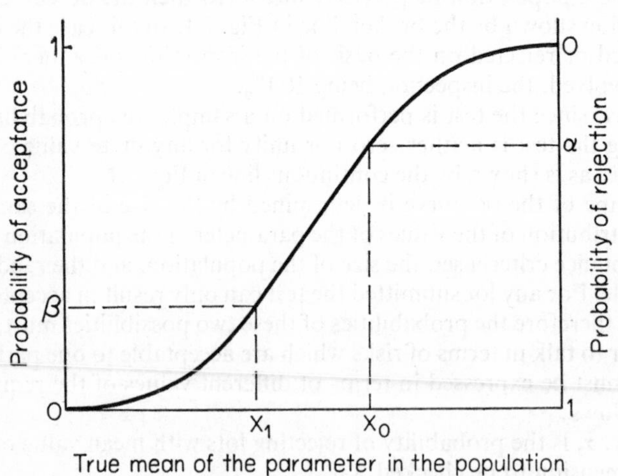

Fig. 7.2 Operating characteristic of a sampling plan showing supplier's and customer's risks

Discrimination ratio of the sampling plan is the ratio of X_0 to X_1. (Depending upon whether more favourable values of X are greater or smaller than less favourable values then the ratio is usually taken as X_0/X_1 or X_1/X_0 whichever is greater than unity.)

X_0 is often called the "specified" value of the required parameter or dimension, and X_1 the "minimum" (or maximum if X is regarded as less favourable when increasing) acceptable value. In some cases the producer's risk, α, may be defined as the probability of rejecting lots with an acceptable proportion of the lot having values worse than specified. Similarly the consumer's risk, β, would be defined as the probability of accepting lots with an unacceptable proportion of the lot having values worse than specified.

The reader will appreciate that the flatter the OC curve, and hence the less effective the sampling (the smaller the sample), the more the supplier must aim to exceed the customer's requirements in order to satisfy his chosen producer's risk, as can be seen from Fig. 7.2.

Typical values of producer's and consumer's risks are in the range 5% to 15%. Larger values tend to be unacceptable and smaller values require either a larger sample (this has to be paid for) or a larger discrimination ratio, which means that the supplier has to design to an even more favourable value of X in order to satisfy the risks (this also has to be paid for). If $\alpha = \beta = 10\%$ then the test can be expressed by saying that, if the value of the parameter in the product is as good as X_0, there is a 90% chance of its passing the test, whereas if it falls as low as X_1 there is a 90% chance of its being rejected.

Since we are concerned with maintainability, the parameter X is expressed in terms of repair times and down times. The following extract from US Military Standard 471 defines some of the terms used:

"B10.3 Identification of symbols

$\bar{M}_{ct}$ = Mean corrective maintenance downtime
$\bar{M}_{pt}$ = Mean preventive maintenance downtime
$\bar{M}$ = Mean maintenance downtime consisting of $\bar{M}_{ct}$ and $\bar{M}_{pt}$ in the same time period

"The mean maintenance downtime is the summation of all maintenance downtimes during a given period of time divided by the number of maintenance tasks during the same period of time.

$M_{max\,ct}$ = Maximum corrective maintenance downtime
$M_{max\,pt}$ = Maximum preventive maintenance downtime

"The maximum corrective or preventive maintenance downtime is that value below which a specified per cent of all maintenance tasks can be expected to be completed. Unless otherwise specified, this value is taken at the 95th percentile point of the distribution of downtimes.

$\tilde{M}_{ct}$ = Median corrective maintenance downtime
$\tilde{M}_{pt}$ = Median preventive maintenance downtime

"The median is that value which divides all the downtime values so that one-half of the values are equal to or less than the median and one-half of the values are equal to or greater than the median."

Military Standard 471 lists a number of procedures for carrying out maintainability tests. Method 1, for instance, assess both $\bar{M}_{ct}$ (mean corrective maintenance time) and $M_{max\,ct}$ at 90 or 95 percentile (the value of corrective time not exceeded by 90 or 95% of the individual values in the population).

Table IV of Military Standard 471 must be studied carefully if the supplier is to understand his true risks. The sequential test, for $M_{max\,ct}$ at 95 percentile, rejects at 4 instances of repair times exceeding the given value for a sample size between 70 and 100 repairs. If a log normal distribution of repair times is assumed then the customer has a 10% risk of accepting for a repair time $M_{max\,ct}$ at the 95 percentile, whereas the unfortunate supplier has a 10% risk of rejection even if he is achieving the required repair time at the 99 percentile. The risk to the consumer of accepting an equipment in which 39% of the repair activities exceed the quoted mean repair time is 6%, whereas if this quoted mean is exceeded on only 22% of the repair occasions then the producer has a 6% risk of rejection. Should the seller, then, assume that the quoted values of $\bar{M}_{ct}$ and $M_{max\,ct}$ are those for which he should aim, his commercial future is likely to be unprofitable.

The mathematics used in Military Standard 471 usually assume a log normal distribution of repair times, and it is stated that this is typical. This may be true of systems employing consistent technologies such as computer and data systems, telecommunications line transmission or telephone switching systems, but complex equipments using a number of different technologies such as is the case with radar systems, aircraft flight controls, radio guidance systems and so on are likely to produce multimodal distributions. This was discussed in Chapter 2 (see Fig. 2.2). One "hump" of the distribution may correspond to repairs in areas where replicated elements are fully monitored and are repaired by speedy plug-in substitution, whereas another may relate to repairs in areas where the elements are in series reliability, where faults are located only after careful checks using specialized external instrumentation, and where replacements require careful on-site adjustment, calibration and checkout. These types of distributions do not lend themselves to easy statistical analysis. Under these circumstances it would be wise to attempt to separate the faults into two or more categories as suggested above and to perform any statistical calculations or tests separately for each group.

If the maintainability demonstration is carried out under artificial conditions rather than waiting for real faults to occur in the field, then the simulated faults used for the test will be "solid" rather than intermittent. Since real-life faults are so often of the intermittent variety the simulated fault test may result in faster diagnosis to the advantage of the producer. Furthermore, the repairs in an artificial test will be carried out by a maintenance man who is under observation and in unnatural conditions. His reaction to this state of affairs might be affected by personal factors which could alter his performance one way or the other. Even assuming that he is not biased in favour of the producer or customer, it is possible that the degree of observation may either inspire him to give of his best or detract from his performance.

The distribution of simulated faults in an artificial sample should be determined by means of the reliability prediction using known failure rates, or by means of a reliability assessment from tests or experience. Any inaccuracy in the reliability assumptions will unbalance the sample and hence distort the maintainability test. Table I, from Military Standard 471, shows how reliability data are used to apportion the sample tasks amongst the various fault areas. The predicted failure rate for each component type is entered in column B, and the quantity of that type in column C. Column D shows the total failure rate contribution for each type, and in column E each of these figures is expressed as a percentage of the total. Column F permits the grouping of some of the figures in column E where values of less than 2% are involved. In this example, for a sample of 100 corrective maintenance actions, 37 would be chosen from transistor failures, 3 would be capacitor failures, 21 would involve resistors, and so on. The procedure for constructing a sample of preventive maintenance tasks is similar except that column B would contain the frequency of occurrence of each task instead of a failure rate. This procedure is described in Appendix A of the Standard.

Appendix B outlines six alternative test methods for demonstrating the various maintainability parameters $\bar{M}_{ct}$, $\bar{M}$, $M_{max\,ct}$, etc.

Test Method 1

This method makes use of the sequential test. The test described earlier in this chapter consists of a fixed sample size and an accept/reject criterion of a certain number of excessive maintenance times. In the sequential test the repairs are carried out one by one, and after each repair the test plan states an accept and a reject criterion. If the number of unacceptable repair times at that point in the test is equal to or greater than the reject criterion the test has been failed. If the number is equal to or less than the accept criterion the test has been passed, but if the number lies between these two figures then the test continues and

MIL-STD-471 **Table I** *Example*

Corrective Maintenance Sample Selection

A	B	C	D	E	F
Item	*Failures/item/ 1000 hours* (λ_i)	*Quantity of each item* (n_i)	*Failures per 1000 hours* $(\lambda_i)(n_i)$	*Item per cent contribution to total mainten-ance tasks*	*Column* E *regrouped*
Category I					
Gears					
Spur	0·0001	10	0·001	0·05	
Worm	0·0001	3	0·0003	0·02	0·09
Helical	0·0003	1	0·0003	0·02	
Category II					
Clutch Facing					
Type A	0·023	3	0·069	3·68	3·68
Type B	0·033	2	0·066	3·52	3·52
Type C	0·005	40	0·200	10·70	10·70
Category III					
Module Type (Flip-flop)	0·0057	25	0·1425	7·59	7·59
Module Type (Audio amplifier)	0·0082	30	0·2460	13·10	13·10
Category IV					
Resistors	0·0002	2000	0·400	21·30	21·30
Capacitors	0·0001	500	0·50	2·66	2·66
Transistors	0·0007	1000	0·700	37·30	37·30
			1·8751		

another repair action is carried out. Such a test is constructed using statistical theory similar to that of the fixed sample test, but it has been shown that in the long run sequential statistical tests yield results slightly quicker than fixed plans. The following shows a small part of the test plan given in Method 1 of Military Standard 471.

Cumulative No. of tasks	*No. of observations exceeding $\bar{M}_{ct}$*	
	Accept	*Reject*
5	0	5
6	0	6
7	0	6
8	0	6
9	0	7
10	0	7
11	0	7
12	0	7
13	0	8
14	0	8
15	1	8
16	1	9
17	1	9
18	1	9
19	2	9
20	2	10
21	2	10
22	3	10
23	3	11
24	3	11
25	4	11

Method 1 tests $\bar{M}_{ct}$ and $M_{max\,ct}$ and an accept decision is only reached when both parameters have been accepted. A log normal distribution of repair times is assumed throughout the test.

Test Method 2

Method 2 tests $\bar{M}_{ct}$, $\bar{M}_{pt}$ and $\bar{M}$ with no assumptions regarding the distribution of times provided that the sample size exceeds 50. $M_{max\,ct}$ is also included in the test, but with the assumption of log normal distribution. This is not a sequential test, and formulae are given for the computation of accept and reject criteria for the desired risks.

Test Method 3
This method is used for demonstrating the median equipment repair time with a sample size of 20 and under the assumption of log normal repair times. This is a fixed-sample-size test, and formulae are given for computing the criteria. The consumer's risk, β, is 10% when the repair times are 2·73 times the specified median and the supplier's risk, α, is 5% at the specified median repair time.

Test Method 4
Method 4 is a test of $\tilde{M}_{ct}$, $\tilde{M}_{pt}$, $M_{max\,ct}$ and $M_{max\,pt}$ for an unknown distribution of repair times using a sample size of 50. $M_{max\,ct}$ and $M_{max\,pt}$ are tested at the 95 percentile; in other words the value which will not be exceeded on 95% of occasions. Supplier's risks are not quoted, and these would have to be calculated from the appropriate statistical theory.

Test Method 5
This is a procedure for estimating, at a given confidence level, the percentage of maintenance tasks in the population which will lie between the maximum and minimum values observed in the sample. For a given confidence level and a given percentage required, there will be a sample size which satisfies these values. A table of sample sizes for values of confidence and percentage is given in the Standard. For example a sample size of 38 ensures that 90% of the tasks will fall between the maximum and minimum limits observed in the test at a 90% confidence level.

No distribution of repair times is assumed. If the method were used as a demonstration method then the operating characteristic would have to be derived.

Test Method 6
Only preventive maintenance tasks are considered in Method 6, and it requires that all possible tasks are carried out for the demonstration. This is likely, therefore, to be rather expensive. $\bar{M}_{pt}$ and/or $M_{max\,pt}$ (at some chosen percentile) are obtained from the test data. The mean is obtained in the usual way, and the value of $M_{max\,pt}$, by ranking the repair times from longest to shortest and obtaining the value by observing the required percentage. Since all the possible maintenance tasks are carried out in the test, the only variable factor is the maintenance technician. Human factors can never be disregarded in any of these tests, but it would seem that a common-sense approach to minimizing or balancing out their effects is likely to prove more effective than statistical reasoning.

DATA COLLECTION

It would be wasteful to regard the demonstration test as no more than a tool for determining compliance or otherwise with some contractual agreement. Each repair carried out is a source of design evaluation and can generate feedback of a practical nature which can be used to modify and improve the equipment. In order for this feedback to be harnessed it is necessary to document each repair activity in the same detail that is desirable in field reporting. Information should be recorded as to the duration of the individual elements of repair time as discussed in Chapter 2 (diagnosis, access, etc.). Spares, tools and test equipment used are also of interest, as are comments from the technician as to his thoughts during the diagnosis of the fault.

Logistic and administrative delays should also be recorded, but, since it is the equipment design which is being tested, they must be excluded from the calculations in the demonstration test.

In conclusion let it be emphasized that in any maintainability, reliability or quality demonstration test the details of the test should be fully described in order to minimize the possibility of dispute over its outcome; furthermore, both parties should be aware of and understand the implications of their statistical risks.

Summary

A demonstration of maintainability involves a *sample* of maintenance actions; therefore both parties incur statistical risks of an unfavourable outcome to the test due only to sampling error.

A demonstration test may be based on real or simulated faults.

Due to the statistical risks involved the producer must aim for a higher level of maintainability than that for which he seeks to gain approval.

The only published standard at present is US Military Standard 471. This outlines the procedure for constructing a representative sample of corrective or preventive maintenance tasks. It gives six test methods each designed for the measurement of one or more repair time parameters ($\bar{M}_{ct}$, $\bar{M}$, $M_{max\,ct}$, and so on).

Data generated in the test should be recorded and fed back to the designer.

Demonstration tests, which should be concerned only with active repair times, should be adequately described in the contract document.

8

Maintainability and Reliability in Contracts

Since the late 1950s in the United States of America quantified statements of reliability objectives and, latterly, maintainability have appeared in both military and civil engineering contracts. These contracts involve penalties for failure to meet these objectives. Recently, in Britain, suppliers of military and civil electronic and telecommunication equipments have found that clauses specifying reliability and maintainability are being included in invitations to tender and in the subsequent contracts. Suppliers of highly reliable and maintainable equipment are often well able to satisfy such conditions with no additional design or manufacturing effort, but incur difficulty and expense since a formal quantification or demonstration of these parameters has never been attempted and the techniques for such a study are unfamiliar. In addition, a failure-reporting procedure may not exist and hence historical data as to a product's reliability or repair time may be unobtainable. The addition of system-effectiveness clauses to a contract therefore involves the suppliers of both good and poor equipment in additional activities.

System-effectiveness contractual clauses range from a few words—merely specifying a failure rate or MTBF together with a statement of which part of the system to which the parameter applies—to some two or three dozen pages containing details of design and test procedures, methods of collecting failure data, methods of demonstrating reliability and maintainability parameters, limitations on sources of components, limits to size and cost of test equipment, and so on. Two types of pitfall arise from such contracts:

1. Those due to the omission of essential conditions or definitions.
2. Those due to inadequately worded conditions which present ambiguities, concealed risks, eventualities unforeseen by both parties, etc.

The following headings are essential if maintainability or reliability is to be specified.

Definitions

If a mean time to repair is to be specified then the meaning of *repair time* must be made clear. Does it refer to total down time, time during which revenue is lost, passive or active elements or both?

Failure itself has also to be defined both at system and subsystem level. It may be necessary to define more than one type of failure (e.g. complete system failure, degradation of performance) or failures for different operating modes (equipment used for reception of Morse code, equipment used for the reception of speech signals) in order to cope with a number of different requirements.

Failure rates, mean times to fail and mean times between failures also require definition as to the meaning of "failure" and "time"—what types of

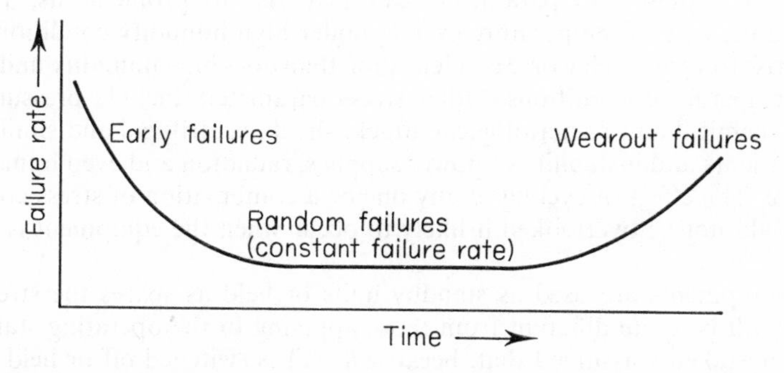

Fig. 8.1 Bathtub curve

failure do not count against the supplier (e.g. maintenance induced by the environment going outside limits), whether time refers to operating time, revenue earning time, real time and so on. Fig. 8.1 shows the traditional *bathtub curve* with the regions of early, random and wearout failures indicated. Reliability parameters usually refer to failures from the random or constant failure rate period, and it is assumed that early failures are removed by burn-in techniques and that wearout failures are eliminated by preventive replacement. It should be made clear in the specification which type of failure is being referred to, and the supplier should take care that he is not accepting responsibility for other types of failure. Parameters should not be specified without regard to their meaning. Failure rate, for instance, has little meaning except when referring to a constant failure rate situation. In a variable failure rate situation an average failure rate might be referred to, but this, if achieved, would only apply over the period stated and might not be satisfied at every point in time. In the case of systems involving redundancy, constant failure rate may not apply despite the use of constant failure rate devices throughout, and either MTBF or reliability should be specified in preference.

Maintainability and reliability are often combined by specifying *availability*, which can be defined in various ways and should therefore be defined in the contract. A common definition is that of the steady-state availability, which is given as MTBF/(MTBF + MTTR), where the MTBF excludes the MTTR.

Environment

A common mistake is to fail to specify the environmental conditions under which the product being supplied is to work. The specification is often confined to temperature range and, perhaps, maximum humidity. This is not necessarily adequate. Even these two parameters can give rise to problems as, for example, in the case of temperature cycling under high humidity conditions. This gives rise to very much worse condensation than does high humidity under constant temperature conditions. Other stress parameters include pressure, vibration, chemical and bacteriological attack, shock, centrifugal and similar forces, transients and instability of power supplies, radiation and even human interference. The effect of cycling of any one or a combination of stress conditions should not be overlooked if likely to occur when the equipment is in service.

Where equipments are used as standby units or held as spares the stress conditions will be quite different from those applying to the operating state. It is often mistakenly assumed that, because a unit is switched off or held in store, it will not fail whilst in that environment. On the contrary, self-generated heat and mechanical self-cleaning actions are sometimes essential to the continued reliability of an equipment. If equipment is transported whilst the supplier is still liable, the very severe mechanical and perhaps temperature conditions which may be encountered must be considered in the contract conditions and in the specifications.

Maintainability is also influenced by environmental conditions. Since repair is nearly always carried out by human beings, environmental conditions pertinent to safety, comfort, health or simply ergonomic efficiency will influence the repair times. The use of protective clothing or remote-handling devices only serves to increase the active elements of repair time by slowing down the technician.

Maintenance Support

The provision of spares, test equipment, manpower, transport and the routine and corrective maintenance of both spares and test equipment is a responsibility which may be divided between supplier and customer or fall entirely on one or other party. These responsibilities must be described in the contract, and the supplier must beware of the risks he runs should the customer fail to meet his responsibilities.

If the supplier is to be made responsible for the training of customer's maintenance staff then levels of intake as well as training have to be outlined in terms of visual acuity, physical ability, reasoning ability expressed in terms of cue and response, manual dexterity and so on, which should be defined in terms of achievement of standard objective tests.

The maintenance philosophy, usually under customer control, will play a part in determining the reliability achieved. Periodic inspection of a non-attended equipment at which failed redundant units are replaced will yield a value of reliability different from the case of immediate repair of all failed units irrespective of whether these failures result in system failures. The maintenance approach must therefore be defined.

Demonstration

The supplier might be called upon to give a demonstration of either reliability or repair time by means of a statistical test. In the case of maintainability a number of corrective repair actions may be demanded, on induced faults of a particular nature, and a given repair time required not to be exceeded for a given proportion of the attempts. In such a situation it is essential to define the type, skill and training of the repair technician, the tools and equipment to be used, the maintenance instructions to be made available and the environment for the test. The method of task selection, the spares to be made available and the repair level must also be stated. The probability of failing the test should be evaluated since some quoted tests carry high risks for suppliers. When reliability is to be demonstrated then a given number of hours may be accumulated and a number of failures stated above which the test is failed. Again statistical risks apply and the supplier must calculate the chance of failing the test with good equipment as must the customer his risk of accepting inadequate goods.* Essential parameters to define here are environmental conditions, allowable failures (e.g. maintenance induced), operating mode, preventive maintenance policy, burn-in or pre-stressing, costs of failures during the demonstration, and cost of operating the equipment. It is often not possible to construct a statistical test of reliability which combines sensible risks for both parties together with a reasonably short duration of test. Under these circumstances the acceptance of reliability may have to be on the basis of accumulated operating hours on previously installed systems belonging to the same or some other customer.

* See Chapter 7 herein and Smith, D. J., *Reliability Engineering* (Pitman, 1972).

An alternative to statistical or historical demonstrations of repair time and maintainability is the existence of a guarantee period wherein all or part of the cost of failure, repair and sometimes redesign work is borne by the supplier. In these cases great care must be taken to calculate the likely costs involved in such an agreement. It must be remembered that, if 100 equipments meet their stated MTBF under random failure conditions then, after an operating time equal to one MTBF, 63 of these systems will, on the average, have failed. Many profit margins have been absorbed by the unbudgeted penalty maintenance incurred due to this fact.

Liability

It is here that the exact nature of the supplier's liability must be spelt out, including the maximum penalty which can be incurred. If some qualifying or guarantee period is to be observed, it must be stated when the supplier will cease to be liable for failures. It is wise to establish a mutually acceptable means of arbitration should the occasion arise where the extent or even the existence of liability is in dispute. If part of the liability for failure or repairability is to fall on some other supplier or a subcontractor, or on a component-part supplier, care must be taken in defining each party's liability. The interface between equipments guaranteed by different suppliers and the case of failures induced in one supplier's equipment due to the degradation in performance or even failure of another's are a potential area of disagreement.

It has been stated that the foregoing paragraphs outline the essential contents of a system-effectiveness contractual agreement. In addition to these, however, the following areas may well be required by the customer to be included in the tender or contract.

Information

Parts list. This may require the type, quantity and supplier to be stated. Provision should then be made for alternative approved sources of supply in case the stated source becomes unreliable or ceases to exist.

Maintainability or reliability analysis. The supplier may be required to offer a detailed prediction of maintainability and/or reliability together with a statement of the procedures and techniques by which it is intended to achieve these objectives. Failure rate data sources may be quoted or sometimes actual data may be given in the invitation to tender and the supplier required to use these data in his predictions. Such data should be compared with the supplier's usual source to establish whether it is optimistic or pessimistic. The comparison should be made overall and between individual component types, for although one component may appear less reliable in comparative data, another may appear to be more reliable. Insistence on optimistic data makes it more difficult to achieve the predicted values, whereas insistence on pessimistic data makes it more likely that unnecessary measures will be demanded to achieve the objectives.

Maintainability and Reliability Programme

The following design, manufacturing and installation programme activities may be required to be outlined.

Tests. Details of screening tests, burn-in tests, qualification and environmental tests.

Costs of maintainability and reliability activities.

Design review. Details of timing and representation on design review committees.

Failure reporting. Details of failure reporting procedures and documentation to be used throughout laboratory, prototype and production model testing.

Failure mode and effect analysis.

Maintenance task analysis.

Built-in test equipment analysis. Including space, weight and power requirements, probability of defect detection and false alarm incidence.

Tools and external test equipment. Details of requirements together with preventive and corrective maintenance of test gear. Details of reliability, spares requirements, weight, volume and power consumption.

Preventive Maintenance Details

A schedule of necessary preventive maintenance may be required to be specified together with a guarantee that the mean and maximum preventive maintenance times do not exceed stated values. The maximum number of preventive maintenance actions in a given time may be subject to some limit, and a description of the effect of each operation on the normal operation of the system may be required.

Other Factors

Storage. The equipment may be purchased by the customer and stored for some time and under conditions different from those of normal operation before it is used. If the equipment is guaranteed for some length of time then it must be made clear what are the storage conditions and for how long these are acceptable. The same applies to the storage of spares and of test equipment and also to the transportation of systems, spares, test gear, etc.

Safety. Hazards to be spelt out together with intended means of protection.

Manuals. Details of operating and maintenance instructions to be provided. The type of fault-finding manual required may be specified (see Chapter 9).

LRA. Limitations on the size, cost and weight of the least replaceable assembly may be imposed.

Repair technicians. The maximum number of repair technicians to work on a given maintenance action may be limited.

Pitfalls

The foregoing lists the possible areas of maintainability and reliability likely to be required to be mentioned in an invitation to tender or in a contract. There are pitfalls associated with the omission and inadequate definition of most of these factors and some of the more serious of these are outlined below.

Repair time. It was seen in Chapter 2 that repair times could be divided into elements. The first division could be into active and passive repair elements, and it appeared that, broadly speaking, active elements were dictated by system design and passive elements by the arrangements adopted by the system user. For this reason the supplier, when committing himself to repair time objectives, must ensure that he is not guaranteeing a part of the repair activity which is determined solely by the customer.

Statistical. A statistical maintainability test is defined in terms of a number of sample repair actions and an objective repair time which must not be exceeded on more than a given number of occasions. Similarly a reliability test is described in terms of a number of test hours during which a number of failures must not be exceeded. In both cases it is possible to calculate the actual MTTR or MTBF which is required to satisfy the test, given that the supplier runs a risk (probability) of failing the test with adequate goods and the customer a risk of accepting bad goods due to an unrepresentative sample. Unless the risks are calculated the parties to such a test may find that they are inadvertently accepting quite unrealistic risks. Where published test plans are quoted for the purpose of specifying a test it is never a bad thing to calculate the risks for oneself. It is not difficult to construct a test that requires the supplier to design equipment 50 times as reliable as is proved in order to pass.

Definitions. The most likely area of dispute is the definition of what constitutes a failure. There are levels of failure (system, redundant unit, spare, etc.), types of failure (performance degradation, catastrophic, wearout, etc.), causes of failure (random, systematic, maintenance induced, overstress, etc.), and also there are effects of failure (dangerous, dormant, etc.). Different MTBF and MTTR objectives and different penalties may apply to each of the many combinations of these factors. It is seldom sufficient, then, to define failure as the inability to meet the performance specification, since some failures may be the fault of the system user and, on the other hand, some failures although the fault of the supplier may not result in an immediate system failure. Careful definition of the various types of failure to be covered by contract conditions is therefore important as well as the definition of failures which are to be discounted.

Quoted specifications. Sometimes a reliability or maintainability programme or test plan is specified as being defined by "such and such a published standard". The standard may be quite applicable to the product in question and the implications to both parties devoid of serious pitfalls; nevertheless, the document referred to should always be studied in detail and the exact extent of the activities required and the risks imposed assessed.

Environment. Environmental conditions affect both repair times and reliability levels. Temperature and humidity are usually quoted, and the pitfall of temperature cycling has already been mentioned. If other factors are likely to influence the design of the product (lightning, power stability, vibration, chemical and biological attack, etc.) then they must either be specifically excluded from the environmental range for which the product is guaranteed or specifically mentioned, in which case they can be designed for and, presumably, allowed for in the price. It is not always desirable to mention every operating stress parameter that can be thought of, since this only leads to additional design effort which will be wasted unless the factor referred to is a real possibility.

Liability. When stating the supplier's liability for failures and maintenance actions it is important that there is no doubt as to the limit of his liability both in terms of cost and time. The supplier must ensure that he knows when he is finally free of liability.

Penalties

There are various ways in which a penalty may be imposed on the supplier to compensate the customer on the basis of maintenance costs or the cost of system outage. The alternatives are briefly outlined below.

Supplier bears costs during demonstration or guarantee period. This method is illustrated in Fig. 8.2(*a*). The supplier pays the total cost of corrective maintenance until the end of the agreed guarantee period. He may also be liable for the cost of redesign made necessary as a result of systematic failures. In some cases the guarantee period recommences if alterations to the equipment are performed or if a given number of failures is exceeded in the guarantee period. The disadvantage of this arrangement is that it offers the customer little incentive to keep maintenance costs to a minimum, since he is responsible for no corrective costs at all until the guarantee period has expired. If the maintenance is actually being carried out by the customer and simply paid for by the supplier, this represents a serious risk for the supplier since there is a

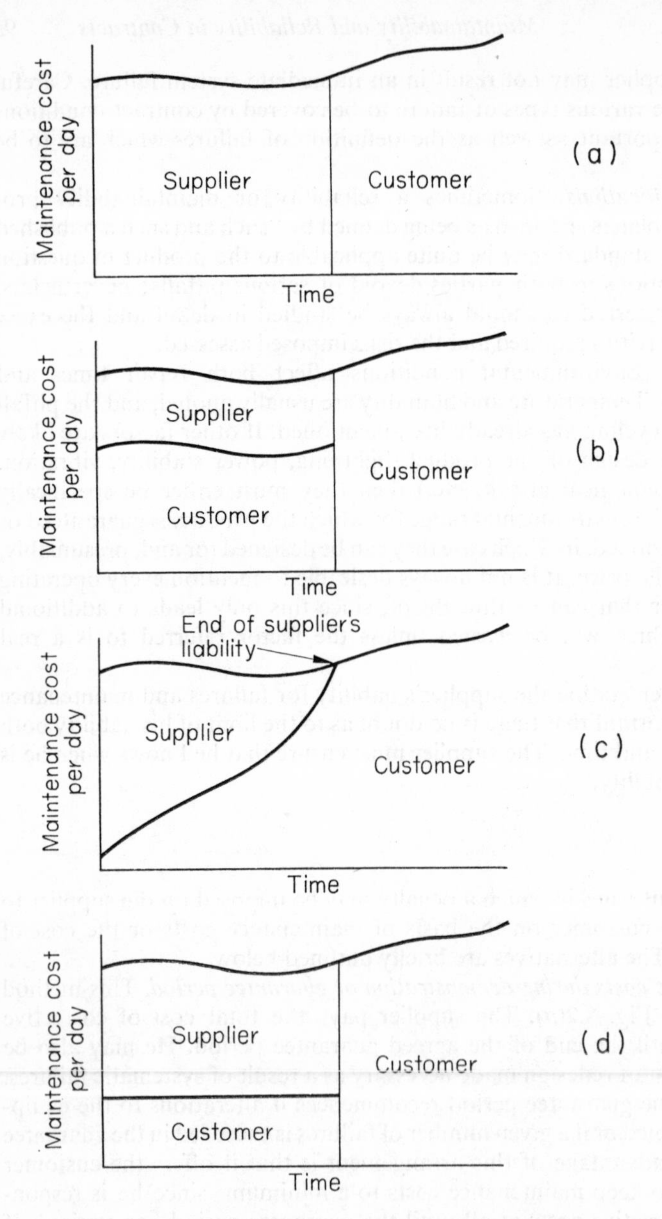

Fig. 8.2 Alternative methods of apportioning maintenance costs between supplier and customer

considerable danger of essential preventive maintenance being neglected in favour of other priorities. The customer should never be permitted to benefit from poor maintenance, and for that reason this method is not very desirable.

Supplier bears proportion of costs during liability period. This is illustrated in Fig. 8.2(*b*). In contrast to the above method this provides both parties with an incentive to keep corrective maintenance costs to a minimum. In the alternative arrangement of Fig. 8.2(*c*), the supplier's proportion of the costs is greater at the beginning but decreases to zero at the end of the liability period. In Fig. 8.2(*d*) the customer's share of maintenance costs remains constant and the supplier pays the excess costs during the liability period. The arrangements at (*b*) and (*c*), by making the customer's share vary with the total costs, provide incentives for both parties. The arrangement at (*d*), however, provides a mixed incentive. The customer has, initially, a very high incentive to maintain the equipment in a reliable state, but once the ceiling has been reached and the supplier is paying the additional maintenance costs then no incentive exists for the customer.

Supplier pays an agreed rate during down time. The above arrangements consist of penalties which are related to maintenance costs. This alternative is related to down time and the cost thereof. Some contracts demand a payment equal to some fixed percentage of the contract price during the down time of the system. Providing that the actual sum paid is less than the total cost of the maintenance then the result is a penalty similar to that illustrated in Fig. 8.2(*d*), except that it is the supplier who pays the fixed contribution. In this case the customer's contribution is still linked to the total cost and he has an incentive to reduce maintenance time.

Except in the case of Fig. 8.2(*a*), it would not be practicable for the supplier to carry out the maintenance for which he was liable. Usually the customer carries out the corrective maintenance and the supplier pays for his share in accordance with some agreed rate. In this case the supplier should require some method of control over the recording of corrective maintenance effort together with a right of inspection of the customer's maintenance records and facilities from time to time.

It should be remembered that achievement of the required reliability and maintainability levels does not mean that no corrective maintenance will take place. If a desired MTBF of 20 000 h is achieved for each of 10 equipments, then in 1 year (8760 h) about 4 failures can be expected. This makes the Fig. 8.2(*a*) method even less desirable and Figs. 8.2(*b*) and (*d*) fairer.

Where a part of a system is to be subcontracted to another supplier then the primary contractor must ensure that he passes on to the subcontractor an appropriate allocation of the system effectiveness commitments in order to protect himself against failure of the subcontracted items.

When negotiating system effectiveness contractual requirements the supplier has an opportunity to ask the customer to agree to his failure reporting demands in return for one or other of the customer's contractual requirements. Field failure and maintenance data are extremely valuable and no opportunity to acquire them should be lost.

Summary

Essential areas to be covered in the contract:

Definitions
Environment
Maintenance support
Demonstration
Liability

Other areas likely to be mentioned:

Parts lists
Maintainability and reliability programme details
Preventive maintenance details
Storage conditions
Safety requirements
Manuals and instructions
LRA
Type of maintenance technician

The following are the major areas of difficulty:

Repair time elements
Statistical risks
Definitions
Quoted specifications
Environment
Liability
Subcontracts

Types of penalty*:

Supplier bears costs during demonstration or guarantee period
Supplier bears proportion of costs during a given period
Supplier pays a fixed rate during down time

Failure reporting:

If the customer demands some maintainability or reliability contractual requirements then the supplier might well make certain failure reporting demands in return.

* See text for further division of these headings.

9

Maintenance Handbooks

The maintenance manual associated with a system should provide all the information required by the user. This may include the following:

Definition of system functions
Statement of system performance
Theory of operation and limitations of use
Techniques and procedures of operation
Permitted range of operating conditions
Supply requirements
Corrective and preventive maintenance routines
Necessary and permissible modifications
Spares provisioning requirements
Spares descriptions and definitions
Test gear definitions, descriptions and check procedure
Disposal instructions, in the case of hazardous materials

The "manual" may range from a simple card to hang on a wall to a small library of information comprising many handbooks for different applications and persons.

The handbook writer is a specialist, and we shall not attempt to specify the information and skills he requires, but we shall outline the essential object of a handbook and discuss some of the methods of presenting the information. Achieved reliability and maintainability depend, to a very large extent, on the quality of the maintenance instructions, and the maintainability engineer has therefore to supply the necessary information to the handbook writer and to collaborate with him if the instructions are to be effective. It is not unknown for handbook writers to draft chapters which contain complete nonsense merely in order to goad the design team into supplying adequate information. This policy, perhaps necessary as a desperate resort, is not an efficient one.

Consider the provision of maintenance information for a complex system operated by a well-managed operating organization with the following maintenance policy.

The system will be maintained, for the most part, by a permanent team (A) who are based on site. This team of technicians, at a fair level of competence, have to service a range of systems and for this reason are not expert in any one particular equipment. Assume that the system incorporates some internal monitoring equipment and some specialized portable test gear for both fault finding and operational checks. This local team carry out all the frequent routine checks and repair most faults by means of modular replacement. There is a local, but limited, stock of certain modules (LRAS), and this stock is replenished from a central depot serving many sites. The depot also stocks replacement items not normally held on site.

Based at the central depot is a small staff of highly skilled specialist technicians (B) who are available to the individual sites. Available to them is further specialized test gear and also basic instruments capable of the necessary tests and measurements likely to be made. These technicians are called into action when the procedures used by the site staff prove inadequate for diagnosis, replacement, checkout or alignment. This team also visit the sites in order to carry out the more difficult or critical maintenance checks and replacements.

Also at the central depot, a workshop staffed with a team of craftsmen and technicians (C) carries out all routine repairs and checkout of faulty modules returned from the field. The specialist team (B) is available for diagnosis and checkout whenever the predetermined routines for module repair and test prove inadequate.

The operations technical headquarters (D) is responsible for the management of the total service operation, including cost control, co-ordination of reliability and maintainability statistics, and the analysis of data, system modifications, service manual updating, spares provisioning, stock control, and, in the event of either loss of service from the original design authority or lack of access to it, a post-design service.

Group A will require detailed and precise instructions for all the preventive and corrective maintenance tasks that they are authorized to carry out. A description of the system function and operation is desirable to encourage interest but must not be so detailed as to encourage unauthorized actions. Instructions for the reporting of all activities must be included, and these should provide for the recording of information which may be of assistance to the diagnosis of failures within the module at the central depot.

Group B will require a far more detailed description of the system and its modules. They require information which will enable them to make a fault diagnosis despite the presence of intermittent, marginal or multiple faults and malfunctions not anticipated at the time when the handbooks were prepared.

Group C will require information similar to that of Group A but concerned with the diagnosis and repair of modules. It may well be the case that certain replacements involve fabrication from raw materials or standard mechanical piece-parts, in which case the necessary drawings and process instructions are required. Some repair techniques may be different from those used in the original manufacture and will therefore need to be fully explained.

Group D require the design data in considerable detail, including some reference to the reasoning used in arriving at design decisions. This may be invaluable in order to arrive at a correct decision for subsequent modifications after many years of service. Detailed spares requirements are essential so that correct and safe substitutions can be made in the event of a spares source becoming unavailable. Consider a large-population item which may have been originally subject to stringent process control and screening for high reliability. Obtaining a further small number of these items to a similar standard may be impossible or at least costly, and it may well be that replacement of a small number of the population with less assured items is acceptable. This may not apply to another item occurring in a few places and critical to the operation of the system. Here the replacement must be of the same high reliability as the replaced item. Consider, also, a component originally supplied to meet a pan-climatic specification. A system user in a particular part of the world may well substitute a locally obtained item which meets his own particular climatic situation although not suitable for universal application.

Only the most fortunate reader will encounter a situation as outlined above with the necessary support to satisfy all requirements. The example, however, indicates some of the factors to be considered in providing maintenance information.

Preventive maintenance procedures will be listed in groups by service intervals. Service intervals can be determined by the calendar, switch-on time, elapsed time, miles travelled, hours flown, number of missions, or number of operations, whichever is appropriate. The procedures and the method of recording the actions must be fully described. In the event of a check giving an unsatisfactory result the corrective action will be described or the user directed to the appropriate part of the corrective maintenance instructions.

In the case of corrective maintenance the likely failure indications must be listed and may take the form of specific malfunction alarms, a fault print-out, failure to pass some routine check, or, in the worst case, noise, smoke, fire or explosion. The last three are, it is to be hoped, so rare as to justify exclusion from the list. The list may take the form of a "fault dictionary". The first actions specified must be those which ensure safety of personnel, and the second must be those which restrict or prevent secondary failures and subsequent damage to the equipment. Then follow the actions which restrict the loss of service to the minimum possible. Certain of these urgent actions may have to be extracted from the manuals and taught as essential security drills or even made the subject of displayed instructions and posters.

The next consideration is the diagnostic procedure followed by the repair and then the alignment and checkout.

Ideally diagnostic and repair processes will be described in a logical flow chart. This will be expressed in terms of "If . . . then . . ., whereas if not . . . then . . .". Fig. 9.1 shows a segment of a typical corrective maintenance algorithm.

Where such a simple process is impossible and the maintenance technician

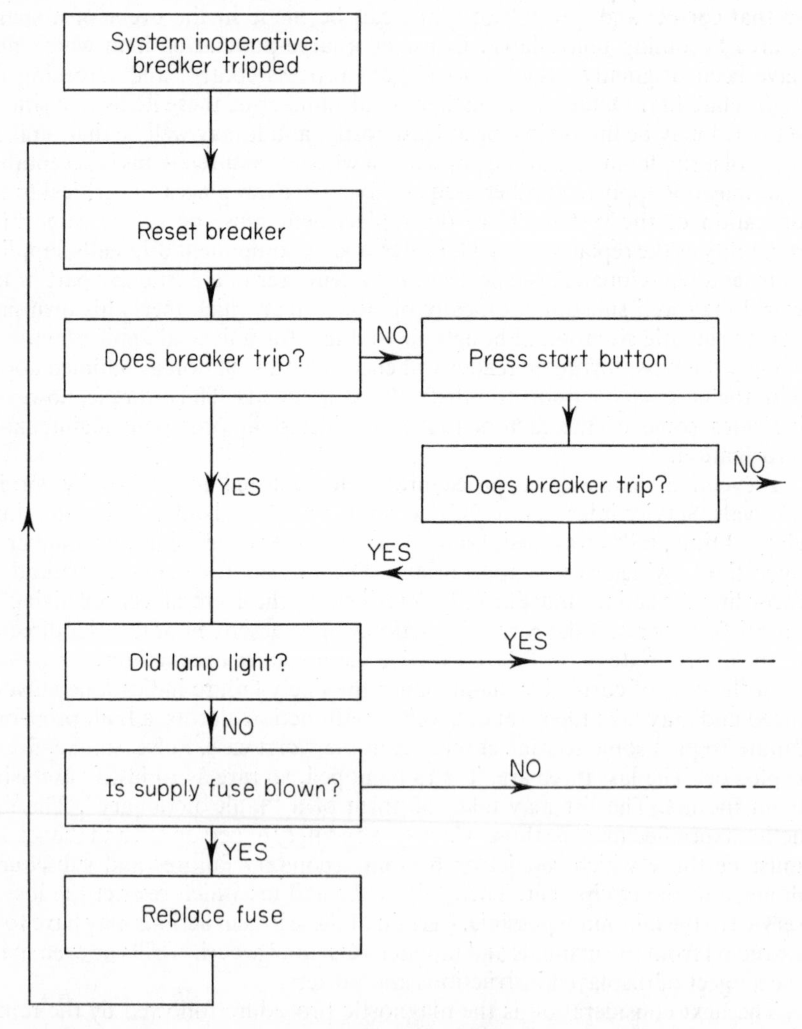

Fig. 9.1 Typical corrective maintenance algorithm

has to follow a logical procedure using his initiative to fill in the gaps in a partially defined sequential process then the presentation of both schematic diagrams and system description becomes very important.

Some faults, by their nature or by their annunciation, indicate the function which is faulty, and therefore the presentation of the information in functional blocks together with an indication of the physical distribution of the functional elements is required. Other faults are best detected by identification of the conditions existing at the interfaces of the physical assemblies, and the information then needs to be presented on a physical element basis with the functions involved indicated. In many cases both approaches will be necessary. This illustrates the desirability, from the maintenance point of view, of designs

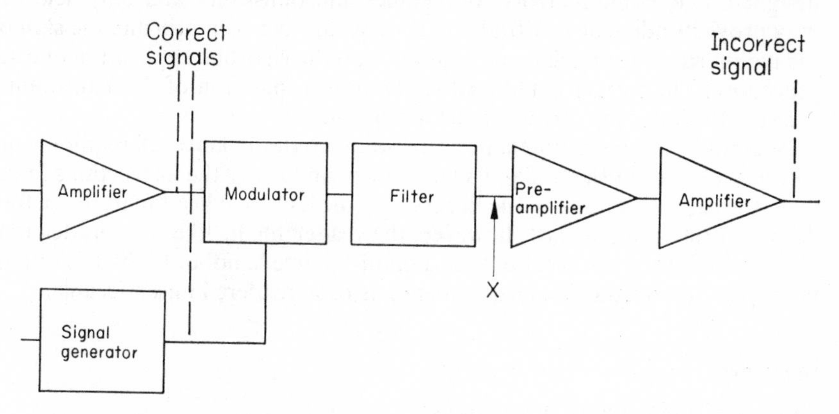

Fig. 9.2 Flow diagram for diagnosis

X. Test signal at this point. If signal is correct, fault is to the right. If signal is incorrect, fault is to the left.

in which the functional and physical realizations are closely related. In a complex system the final location of the fault may be by means of a bracketing elimination process. For example, "The required signal arrives at unit 12 but is not present at unit 20. Does it appear at unit 16? No, but it appears at unit 14. Change unit 15." Fig. 9.2 is an example of this type of reasoning presented in flow diagram form.

Perhaps the most highly developed technique for maintenance documentation is that known in the Royal Navy as FIMS (functionally identified maintenance system). The original concepts were developed by Technical Operations Inc., applied by the United States Coastguards as SIMS (symbolic integrated maintenance system) and in the US Navy as SIMM (symbolic integrated maintenance manual). The British Navy has reflected the concepts back into the development phase employing a Design Documentation System Co-ordinator, who has both system effectiveness and documentation responsibility. This arrangement brings appropriate emphasis to bear on reliability and maintainability in the design stage of a system.

Where maintainability is of paramount importance and where the great diversity of equipments and the stresses of operating conditions render even the most highly skilled almost entirely dependent upon good documentation, the demand for effective documentation reflects back into the design itself to the great benefit of reliability and maintainability.

It must not be assumed that the provision of a handbook automatically guarantees correct diagnosis and repair. The handbook, however carefully designed, may contain errors, ambiguities and omissions, and only field use may prove its adequacy. A trade-off is necessary between utilizing the skill of the maintenance technician and insistence on the rigorous application of a set procedure. The former could lead to the non-recognition of dormant faults, whereas the latter may frustrate the technician.

According to the maintenance environment the balance of requirements will result in one of the trade-offs mentioned, and hence the instructions range from the card on a hook to the library at some technical headquarters. Before despising the humble card, however, the reader might like to consider that a card put to good use is better than a multi-volume handbook which is sought in desperation only after an equipment has been rendered unserviceable.

Summary

Objects of maintenance instructions:

Reduce manhours
Reduce skill level
Reduce mean down time
Reduce incidental damage due to incorrect preventive or corrective maintenance
Ensure correct calibration and checkout

Manuals may range in complexity from very small and simple to very large and complex.

Different manuals may be required for different types of maintenance team, i.e. first line, second line, headquarters, etc.

Methods of event recording should be included in manuals.

Instructions should include a fault dictionary.

Diagnosis may be aided by an algorithm of decisions.

Diagnosis may, alternatively, be aided by a bracketing elimination procedure. This is employed in a Royal Navy system known as FIMS.

Where diagnosis is too complex to be totally described by such methods then full diagrams and system descriptions should be included.

Documentation should be integrated into the design process, and as a result may have some effect on design.

Provision of the handbook does not, in itself, guarantee correct diagnosis. The handbook may require modifications as a result of inadequacies brought out in use.

A manual which is not used is a waste of money.

10

Maintainability in the Future

The supply of operating engineering systems will soon be limited by the resources available to maintain them. Already, serviceable equipments are being thrown away and replaced, because the skills or the information required for repair are not available. As a result the squandering of the earth's resources is increased.

Consider the maintenance organization for some future national or even continental service. The central depot may be fed with a continuous print-out showing

1. System status, i.e. failures existing; any degradation of service; degree of risk of specific degradations; risk of loss of service.
2. Actions required, e.g. no action; repair at next routine visit to site; repair at special visit to site within the next week; take immediate action.

If it is decided to make a visit to a site the technician assembles the necessary spares and equipment and travels to the site. For reasons of urgency or perhaps convenience he may travel by VTOL (vertical take-off and landing) craft, major sites having being selected and equipped for this method of transport. Because repair can be speedy the required availability can be achieved by a combination of short MTTR and equipment redundancy. As a result, no local inactive spares are held except for a few low-cost expendable items and small tools.

The most common form of replacement assembly would, perhaps, be a wheeled trolley having no obvious electrical contacts but with simple mechanical location and latching devices. The signal interface with the main assembly is opto-electronic and power is accepted into ducts, the power supply being in the form of two or three independent low-velocity air jets which can safely be vented into the atmosphere during interchange. The assembly has no fuses or circuit-breakers, and in fact these only occur on the pneumatic power generators and print-out assemblies.

On arrival at the site the local print-out machine provides information with which to verify the latest state of the equipment. The faulty unit is unlatched and wheeled out, the print-out announcing the fact both locally and at the central depot. The replacement unit is wheeled into place and latched, and the print-out then starts to report the new status of the system as it is established by the automatic check-out procedure.

When the defective assembly is opened up at the base workshop the technicians will have, not only a full print-out of its condition produced by an automatic tester, but information from colour changes to built-in electrochemical indicators which identify faulty sub-assemblies or areas where further diagnosis is necessary.

A flight of fancy—yes—but such techniques are known and some are already in use. Systems whose design has been computer aided could well defy fault analysis where comparable computer aid is not also available to the maintenance staff. Under such circumstances the techniques described above could well prove essential to meet system effectiveness objectives.

Glossary

These definitions of terms used in the text are, wherever possible, in line with one or other of the international or national standards.

Maintainability. The probability that a device which has failed will be restored to operational effectiveness within a given period of time when the maintenance action is performed in accordance with prescribed procedures.

Reliability. The probability that a system or device will operate for a given period of time and under given operating conditions.

Availability (Steady state). The proportion of time that a system is available in a very large time interval. (Availability can also be expressed in terms of instantaneous availability—the probability that a system will be available at any random time.)

Down time and repair time. See Chapter 2 for a full discussion of down time, repair time and their elements.

Mean time to repair. The total maintenance time over a number of maintenance operations divided by the total number of operations. Usually this refers specifically to corrective or preventive maintenance.

Failure. The termination of the ability of an item to perform its required function.

Misuse failure. Failure attributable to the application of stresses beyond the stated capacities of the item.

Inherent weakness failure. Failure attributable to weakness inherent in the item itself when subjected to stresses within the stated capabilities of the item.

Sudden failure. Failure that could not be anticipated by prior examination.

Gradual failure. Failure that could be anticipated by prior examination.

Partial failure*. Failure resulting from deviations in characteristics beyond specified limits but not such as to cause complete lack of the required function.

Complete failure*. Failure resulting from deviations in characteristics beyond specified limits such as to cause complete lack of the required function. The limits referred to in this category are special limits specified for this purpose.

Catastrophic failure. Failure which is both sudden and complete.

Degradation failure. Failure which is both gradual and partial.

* The limits set for partial and complete failure must be function dependent in each case. Usually the proportion of failures which might give rise to problems with these definitions is quite small.

Bibliography

GOLDMAN, A. S., and SLATTERY, T. B., *Maintainability—A Major Element of System Effectiveness* (Wiley, 1964).

BLANCHARD, B. S., and LOWERY, E. E. *Maintainability Principles and Practice* (McGraw-Hill, 1969).

BS 4200, *Guide on the Reliability of Electronic Equipment and Parts used therein* (British Standards Institution).

US Military Standard 721B (25.8.66), *Definitions of Effectiveness Terms for Reliability, Maintainability, Human Factors and Safety.*

US Military Standard 781B (15.9.67) with Notice 1 (28.7.69), *Reliability Tests: Exponential Distribution.*

US Military Standard 785A (28.3.69), *Reliability Program for Systems and Equipment Development and Production.*

US Military Standard 1472 (9.2.68), *Human Engineering Design Criteria for Military Systems, Equipment and Facilities.*

US Military Standard 470 (21.3.66), *Maintainability Program Requirements.*

US Military Standard 471 (15.2.66), *Maintainability Demonstration.*

US Military Handbook 472 (24.5.66), *Maintainability Prediction.*

Defence Standard 00–5, Parts 1–3, Issue 1, *Design and Constructional Criteria for Reliability and Maintainability of Land Service Materiel.*

SMITH, D. J., *Reliability Engineering* (Pitman, 1972).

BAZOVSKY, I., *Reliability Theory and Practice* (Prentice-Hall, 1961).

STAFF OF ARINC CORPORATION, *Reliability Engineering* (Prentice-Hall, 1964).

MYERS, R. H., WONG, K. L., and GORDY, H. M., *Reliability Engineering for Electronic Systems* (Wiley).

MORONEY, M. J., *Facts from Figures* (Pelican Books).

WALD, A., *Sequential Analysis* (Wiley).

IEC Publication 271 (1969), *Preliminary List of Basic Terms and Definitions for the Reliability of Electronic Equipment and the Components (or Parts) used therein.*

Index

Plate 1 Remote control unit

(T.522)

Plate 2 Transmitter and monitoring unit

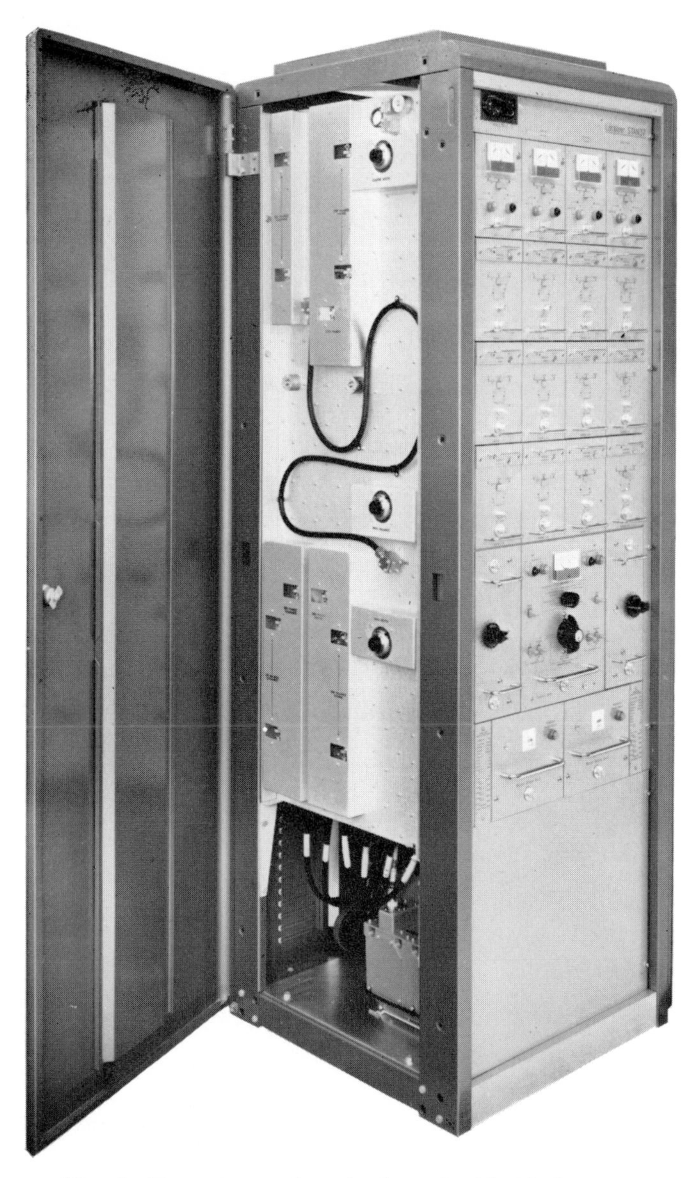

Plate 3 Transmitter and monitoring unit with side door open

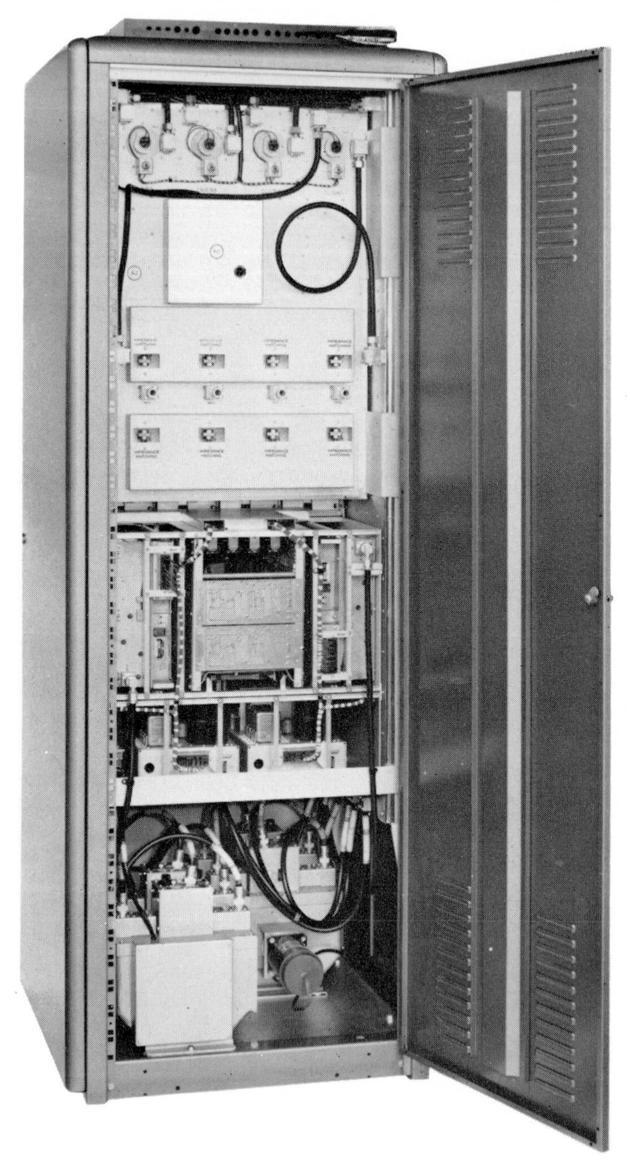

Plate 4 Transmitter and monitoring unit with rear door open

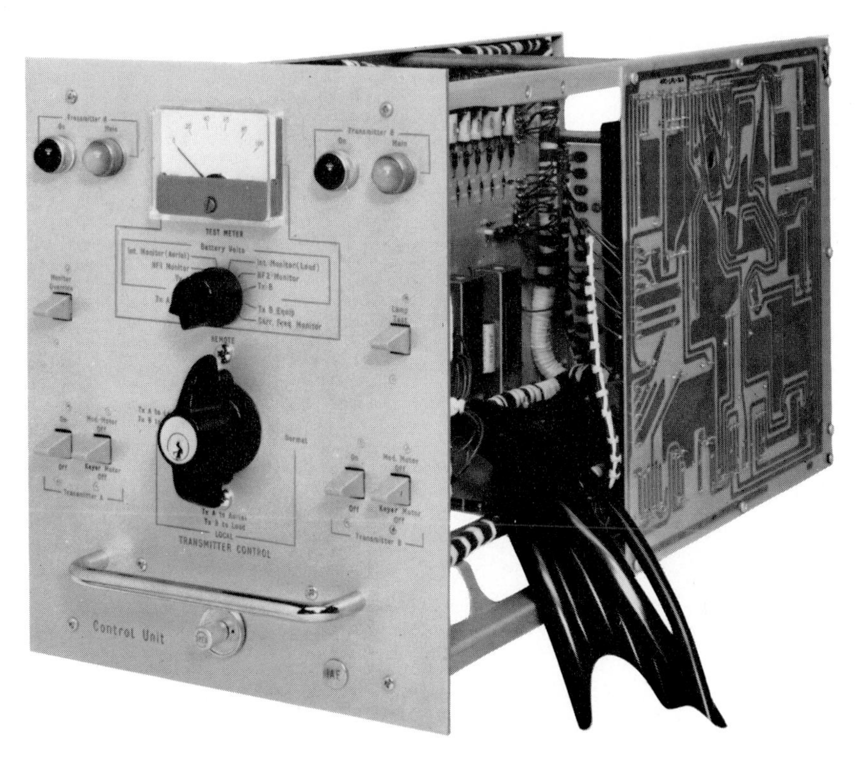

Plate 5 Common control unit

Plate 6 A monitoring unit

Plate 7 An alarm unit

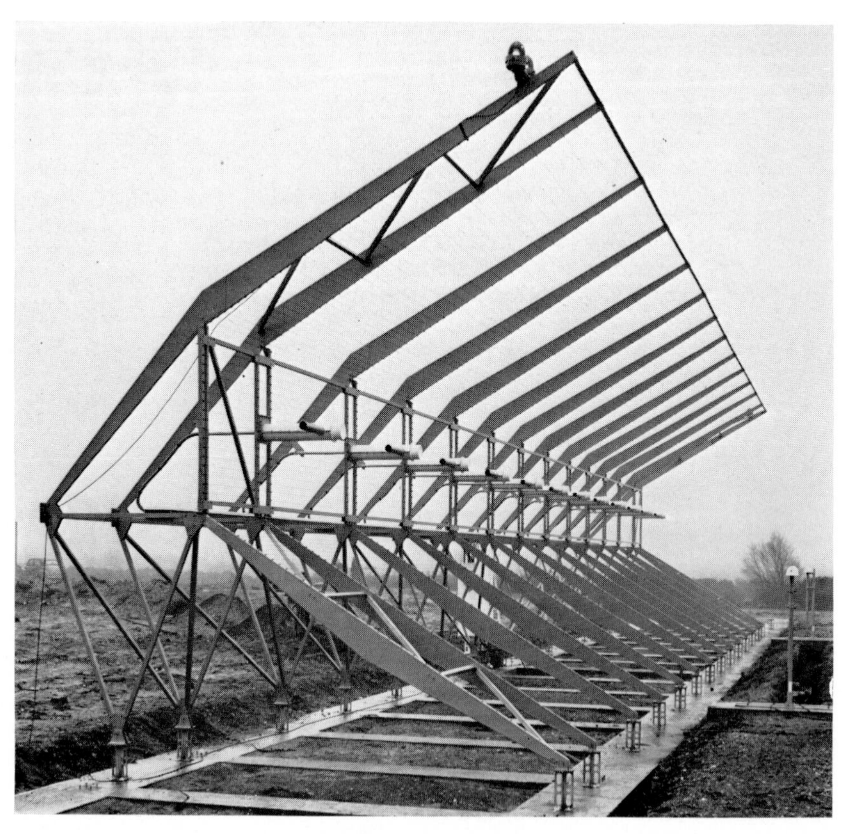

Plate 8 Aerial array